AF270479

MAKING
WEB CONTENT
Essential Library
An Imprint of Abdo Publishing
abdcbooks.com
BY MARIE JASKULKA

Published by Abdo Publishing, a division of ABDO, PO Box 398166, Minneapolis, Minnesota 55439. Copyright © 2025 by Abdo Consulting Group, Inc. International copyrights reserved in all countries. No part of this book may be reproduced in any form without written permission from the publisher. Essential Library™ is a trademark and logo of Abdo Publishing.

Printed in the United States of America, North Mankato, Minnesota.
052024
092024

Cover Photos: Alexey Boldin/Shutterstock Images (computer); Shutterstock Images (mouse); Viktoriia Ablohina/Shutterstock Images (sound wave); Dmytro Zinkevych/Shutterstock Images (phone and light)
Interior Photos: Viktoriia Ablohina/Shutterstock Images, 1 (left); Dmytro Zinkevych/Shutterstock Images, 1 (right); Shutterstock Images, 4, 10, 100, 101; Tiffany Hagler-Geard/Bloomberg/Getty Images, 7; Sheldon Cooper/SOPA Images/LightRocket/Getty Images, 13; Matt Cardy/Getty Images News/Getty Images, 14; Nathaniel Noir/Alamy, 16; Roger Ressmeyer/Corbis Historical/VCG/Getty Images, 19; Martial Trezzini/Keystone/AP Images, 21; David Paul Morris/Getty Images News/Getty Images, 24; Stanislav Kogiku/SOPA Images/LightRocket/Getty Images, 27; Jim Holden/Alamy, 28; Casimiro/Alamy, 32, 45; Harry Murphy/Sportsfile/Getty Images, 35; PjrStudio/Alamy, 37; Mario Martinez/Moment/Getty Images, 38; Miljan Zivkovic/Shutterstock Images, 41; John Patriquin/Portland Press Herald/Getty Images, 42; Gabby Jones/Bloomberg/Getty Images, 46, 63; Michael Kovac/Getty Images for Acura/Getty Images Entertainment/Getty Images, 51; Tali Arbel/AP Images, 55; VCG/Visual China Group/Getty Images, 56; Kelly Defina/Getty Images Sport/Getty Images, 60; Richard Rodriguez/Getty Images for Samsung Galaxy Creators Collective/Getty Images Entertainment/Getty Images, 65; Syazwan Azmi/Shutterstock Images, 66; Mary Altaffer/AP Images, 70, 86; Jacek Sopotnicki/Alamy, 74; Monica Schipper/Getty Images for Nickelodeon/Getty Images Entertainment/Getty Images, 77; Kimberly White/Getty Images for TechCrunch/Getty Images Entertainment/Getty Images, 78; Jacopo Raule/Getty Images Entertainment/Getty Images, 80; Gotham/GC Images/Getty Images, 83; Mike Coppola/Getty Images for YouTube Shorts/Getty Images Entertainment/Getty Images, 89; Kevin Mazur/Getty Images for YouTube/Getty Images Entertainment/Getty Images, 90; Lorenzo Di Cola/NurPhoto/Getty Images, 94; Justin Sullivan/Getty Images News/Getty Images, 98

Editor: Laura Stickney
Series Designer: Maggie Villaume

Library of Congress Control Number: 2023949545

Publisher's Cataloging-in-Publication Data
Names: Jaskulka, Marie, author.
Title: Making web content / by Marie Jaskulka
Description: Minneapolis, Minnesota: Abdo Publishing, 2025 | Series: Making media | Includes online resources and index.
Identifiers: ISBN 9781098293390 (lib. bdg.) | ISBN 9798384912668 (ebook)
Subjects: LCSH: Web sites--Design--Juvenile literature. | Online authorship--Juvenile literature. | User-generated content--Juvenile literature. | Streaming technology (Telecommunications)--Juvenile literature. | Web publishing--Juvenile literature.
Classification: DDC 006.7--dc23

CONTENTS

EVERYONE IS A CONTENT CREATOR

As a web content creator, Malik always had a ton of ideas. That's where his work started. He had to decide if each burst of inspiration was an idea worth pursuing. A lot of work went into publishing a new story on his personal website.

Malik was a digital investigator. Since elementary school, he'd wanted to be a journalist when he grew up. He liked being the first person to learn about a new situation and share the story with the community. Malik had seen journalists on television uncover huge injustices, helping many people in the process. He'd also read breaking news stories online and in newspapers. Malik believed investigating local events and reporting them to the public was good for everyone.

With an eye on his future goals, Malik worked hard to get good graces. He read news feeds on current events sites and had already started looking at colleges with well-respected journalism programs. But because of the World Wide Web, Malik didn't have to wait to start making his own content and sharing it with the public.

People interested in creating web content can explore websites, watch online tutorials, and read articles about web design.

BUILDING A WEBSITE

During his sophomore year of high school, Malik launched a local news blog to practice writing news stories. The first step was choosing a website-building platform that was simple enough for beginners. This would help Malik learn how to build a site. After reviewing different website platforms, he chose Wix. He signed up for the platform's basic account plan. The platform would host Malik's website and store his photos and stories for free.

Once he finished setting up the site, Malik created a posting schedule. He committed to posting one story a week. The first few stories seemed to take him forever to finish, but he stuck to his goal. Over time, Malik learned more about effectively managing a website. Soon, Malik's site began to attract daily visitors. It was getting good web traffic. This refers to the number of people who visit a website, along with how often they visit and how much time they spend there. A site that's getting traffic is doing something right. Having regular readers motivated Malik to work harder on his news stories.

Malik took note of which stories got the most views. Stories about local crimes that never made national or state news seemed to be the most popular. So Malik started

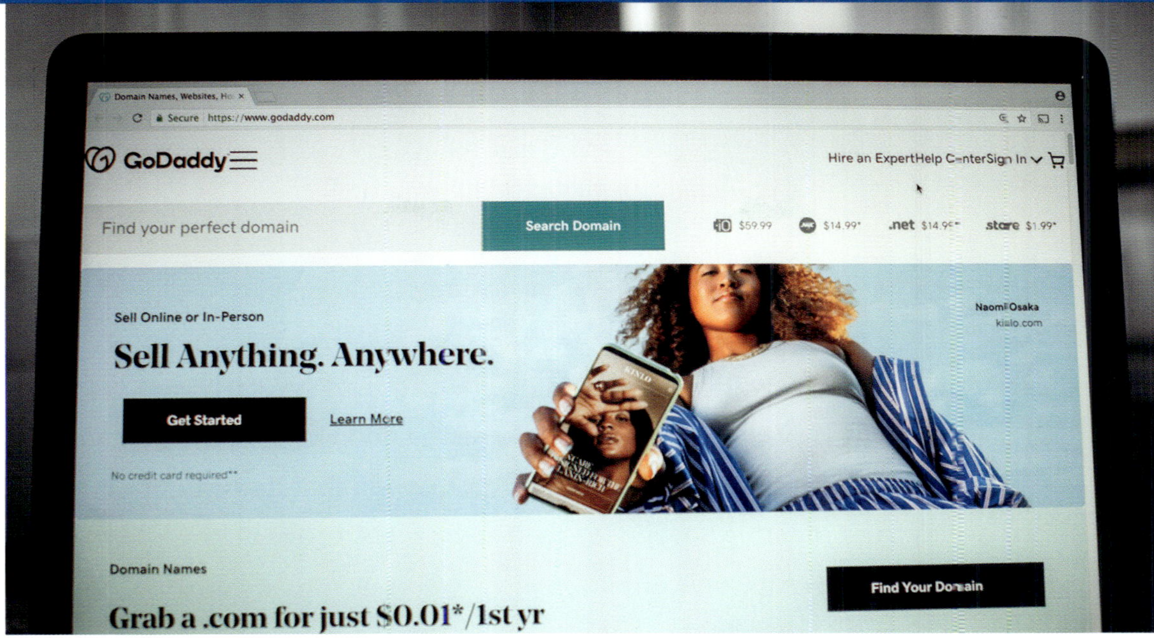

Content creators can choose from many different domain name registrars. One of the largest and most well-known domain registrars is GoDaddy. It manages more than 84 million domains.

writing more content about the topic. Regularly publishing articles helped him hone his reporting skills. He tried to be the first to report a breaking news story before other news outlets. Soon he began posting stories every day.

After a while, Malik decided to switch his site to a different platform. He chose WordPress.org, an open-source software tool that allows people to create their own websites. Users can host the sites themselves, but most people pay someone else to host their sites. To set up his WordPress site, Malik had to purchase a domain name—the name of his website's web address. Malik typed the name of his old blog into the website of a domain name registrar, a company that sells domain names. He was surprised to find that the name was already taken.

Malik needed to come up with a new name. He brainstormed several names before he settled on one that was available. Once Malik chose the domain name, he paid to use it for a year. This meant no one else could use the name during that time period. Malik would also have to pay a monthly hosting fee to keep his website live on the internet. Web-hosting sites provide space on a physical server to securely store a website's data. This way, people all over the world can access the website.

While learning how to use WordPress took some time and money, Malik enjoyed having more control over how his site looked. By watching YouTube videos and reading free tutorials, Malik was able to learn more about designing the kind of site he wanted. Malik's website is one of more than one billion websites on the internet.[1]

WRITING A STORY FOR THE WEB

Just like a professional journalist, Malik knew it was important to always keep his ears and eyes open. He had trained himself to search for news stories everywhere he went. One morning, he went for a run at the local park. As he turned around the bend of the path, Malik noticed that some of the new playground equipment in the park had been vandalized. Pausing the timer on his running app, Malik walked over to inspect the scene more closely. He noticed that the vandals had carved a distinctive symbol into the wood of the slide. He grabbed his phone and took some photographs of the symbol.

Ideas began flowing through Malik's mind. He opened a voice-to-text app on his phone and pressed the Record button. Then he held his phone up to his mouth and spoke into the phone's microphone. The app would transcribe Malik's speech into a text file, which he could then access on his laptop at home.

"It appears the vandals drove into the playground grass and left behind tire tracks," Malik said into the phone's microphone. He snapped some photos of the ruined grass. He peeked at his phone screen to make sure the words had been captured correctly. Then he returned to the path to finish his run, pausing every now and then to record more ideas on his phone.

During his post-run stretches, Malik used his cell phone to search online for any information about vandalism at neighborhood parks. In less than a minute, he found a posting in a community Facebook group. In the post, people complained about damage, which was similar to what Malik had seen on the park equipment. Someone had carved the same symbol into the slide.

Back home, Malik sat down with his laptop. He opened his voice-to-text app and read through the notes he had made on his run. He used them to type up a short article about what he had seen at the park. Malik signed in to his WordPress site and pasted the text into a template he'd created. On his phone, he used another app to brighten and crop the photos he took at the park. Then he added the photos to his blog, typing captions for each picture.

He also added a link to a news story he found that discussed damage at another park. A link is a quick way to get to another website, file, or web page. Content creators use links to connect related pages on the internet. In his article, Malik asked readers to share any information they had about the vandalism with the local police. He also encouraged readers to share his story online. Sharing is the best way to spread web content online. Content may even go viral, getting quickly spread by and shared by millions of people.

SPREADING A STORY

Later, Malik learned he could put ads on his website through the Google AdSense program. Ads are one way that websites earn money. Every time a reader clicked on one of the ads on his site, Malik earned a small commission. A few local businesses asked if Malik would allow them to place ads on his site too. Malik collected an advertising fee from those businesses every month.

People can use voice-to-text apps on their phones to record their ideas on the go. They can also use voice recorders, such as the Voice Memos app that comes preinstalled on iPhones.

Another way Malik attracted new website visitors was through his newsletter. People who visited his site could sign up to join his mailing list. If they signed up, Malik's stories were sent right to their email inboxes. By having a mailing list, Malik connected with people who showed interest in consuming more of his web content.

Every time people saw Malik's name in their inboxes or shared one of his stories, Malik created new pathways in his personal network. This helped him solidify his online identity as a journalist. Collecting email addresses was Malik's way of staying connected to his readers. He hoped a leading news organization might hire him someday, partly because of his robust following. If Malik began writing for a professional news site or newspaper, his readers would likely follow him there. That would create more web traffic for the organization.

Because of this, Malik constantly looked for ways to drive new readers to his site. He used Canva to design a quick video about the vandalism story. He added colorful graphics before posting the video to TikTok. People who saw the video could check out Malik's TikTok profile. His "Bio" page included a link in the comments that took readers to

the original story on his website. While getting likes and follows on social media sites was nice, Malik's main goal was driving people to his website.

WEB CONTENT AFFECTS THE REAL WORLD

By the time Malik arrived at school the next morning, people had begun to read his article. As the day went on, they posted comments about it and shared it with their friends. They added their thoughts about the crime to the discussion. By the end of the school day, people from all over the community had chimed in on the story in the comments section of Malik's website. During a break between classes, Malik scrolled

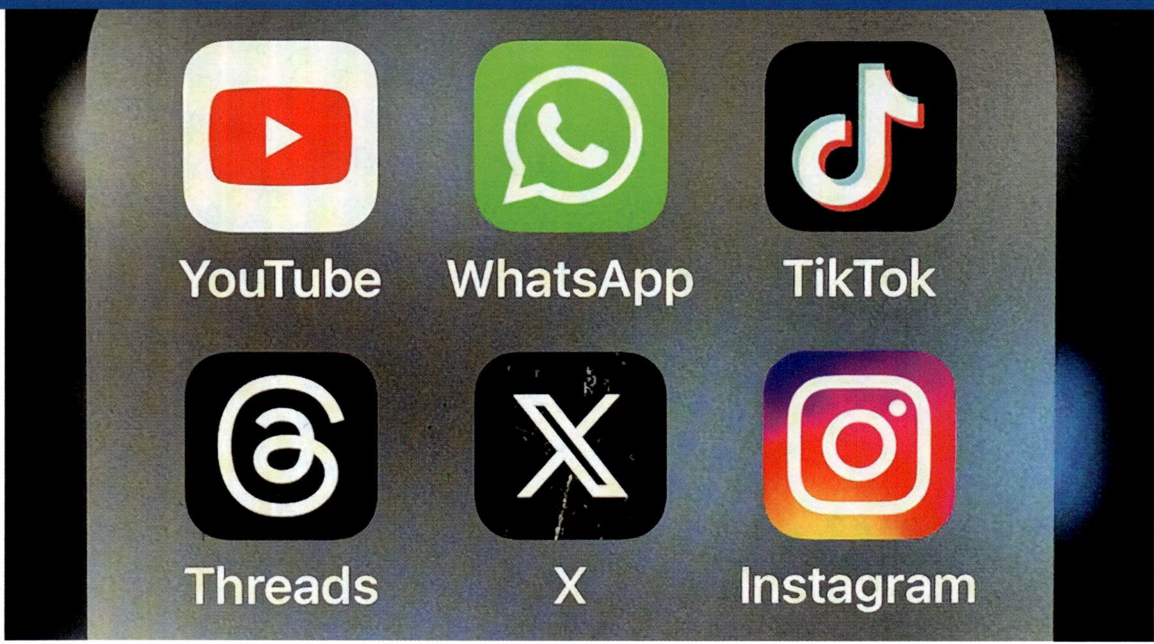

Website creators can reach potential viewers on social media platforms such as Instagram and TikTok. To drive people to their websites, they should post frequently and use colorful visuals.

through all the conversations happening on the site. He couldn't wait to check his stats and web traffic.

When Malik got home from school, he signed in to his website. Anyone can see the pages of a live website. But as the website's creator, Malik could access a "backstage" part of the site. There, he could write, design, and edit his content. In the backend of a website, the creator may also be able to review the site's stats and traffic. Malik was blown away by the number of visitors who had checked out his story—it was more visitors than he had ever had.

The local police had even started investigating the damage at the park. They received phone calls from people who had read Malik's article. Later, Malik turned on the TV to watch the evening news. The TV reporter said the vandals involved in the damage at the park had been arrested. The reporter even mentioned that social media tips had helped police locate the culprits, although they didn't mention Malik's site by name. The prime-time reporters got the big scoop, but Malik knew the story never could have happened without his reporting.

That night, Malik updated his blog, adding a link to the television news footage. Then he went to a private page of his website that only he could access. It was titled "Malik's Journalism Portfolio." He added the story to the long list of stories he'd published since sophomore year. By the time Malik started applying to colleges, he'd have a collection of published stories to share as evidence of his commitment to journalism.

That was Malik's original intention for his website, and he had achieved it. But he had also discovered other benefits of making web content. The ads and subscribers on his website meant Malik didn't have to work a part-time job. His website allowed him to earn money, practice his future career, and make a difference in his community—all in one day.

HELLO WORLD
HELLO WORLD
HELLO WORLD
HELLO WORLD
HELLO WORLD
HELLO WORLD
HELLO WORLD
HELLO WORLD
HELLO WORLD
HELLO WORLD
BREAK IN 20
READY.
PET
2001 Series
personal computer
commodore
TRY YOUR HAND AT CODING
Try typing this in and see what happens!

THE HISTORY OF WEB CONTENT

It may seem like the internet has existed forever. But compared with other mediums such as books or television, the internet is still very new. People have always been inventing new ways to share and learn information. Plenty of people conceived of something like the internet long before it was invented. Scientists, philosophers, mathematicians, and other dreamers came up with ideas that people eventually connected to create the web.

BEFORE THE WORLD WIDE WEB

Before the internet, people had other options to share and consume information. These options still exist today. People can read books, newspapers, magazines, and other printed materials. If people want to learn more about a business, they ask for a brochure or business card. When people want to hear music, they can listen to the radio, hoping the disc jockey plays their favorite song. CDs or records allow them to listen to those songs on their own. People also watch movies on video tapes

One of the earliest personal computers was the Commodore PET 2001. This 1977 computer featured a small keyboard, built-in cassette player, and a memory of four kilobytes.

NETWORKS

and DVDs. Before that, people went to movie theaters every time they wanted to watch a film. However, the internet changed how people consume almost every kind of media.

Today many people have personal computers (PCs) in their homes that connect to the internet. Long before that was possible, the US government used networked computers in the 1960s. Government workers used the network to share secret information without having to use a human messenger or send a letter through the mail. At the time, tensions were rising between the United States and the Soviet Union.

The new technology of connecting computers allowed the US government to share sensitive information on a private network. Government agencies adopted the same technology. Many businesses and universities built their own networks as well, making it easier for employees and scholars to share information. However, none of these individual networks were connected to each other. Outsiders couldn't access a group's network.

Early computers were found mostly in places where people did their work or homework, such as schools. The first PC, the Kenbak-1, was released in 1971. This computer had a small memory storage and a row of switches and lights. But it didn't do much. Only about 40 Kenbak-1s were ever sold.[1]

In the late 1970s, more middle-class people began to buy PCs for their homes. In 1977, three popular PCs were released. The company Commodore released the Commodore PET 2001 computer, which was easy to program and often used in schools. Meanwhile, Apple released the Apple II. This computer was marketed as a device for everyday people. It was celebrated for its simple design and graphics display.

Tandy Radio Shack's TRS-80 Model 1 desktop computer became another favorite for schools and households. Many early computers offered games and art programs. This new technology was fun, but the computers

The Apple II computer was designed by Steve Wozniak. More than two million had been sold by 1984.

weren't connected. When a person finished playing a computer game, they had to go to the store and buy a new one. Still, people loved their home computers. In 1982, *Time* magazine named "The Computer" the Machine of the Year, in place of its traditional Man of the Year.

INFORMATION EXPLOSION

In 1989, British scientist Tim Berners-Lee developed the idea for the World Wide Web, or WWW, to meet the demand of scientists who wanted a faster method for sharing research. Today, many web page addresses still include the letters WWW. The World Wide Web, or web, is different from the internet. It refers to the collection of sites found on the internet. Each web page is a document in a giant web of interrelated information.

In 1990, Berners-Lee and systems engineer Robert Cailliau wrote a paper describing a web of documents that people could view through a browser. Browsers are the software that allow people to look at stored data. The paper explained how users moved from document to document using hypertext. Within a year, Berners-Lee had built the first web server and browser. Berners-Lee asked developers to help him create more browsers.

Their attempts led to the creation of a user-friendly browser called Mosaic. The creators of Mosaic went on to create a browser for PCs, a development that hastened the spread of the web. In 1994, web developers and users gathered at the First International

In 2009, Tim Berners-Lee, *left*, and Robert Cailliau, *right*, celebrated the twentieth anniversary of the World Wide Web.

Conference on the World-Wide Web to discuss their achievements and ideas. The web was a hit.

As more people added pages to the internet, people created businesses to help with connection issues and manage the flow of information. America Online (AOL) was one of the earliest internet service providers (ISPs). It helped many first-time users connect to the Internet through dial-up service. This early method of connecting to the internet was slow and required a phone line, meaning no one else could use the phone number to make phone calls while someone was surfing the web. If someone needed to make a call, the computer user had to disconnect from the internet. Once the phone was done being used, the computer user had to reconnect to the internet.

People also needed something to find their way around the web, which inspired the invention of the search engine. The search engine Jerry and David's Guide to the World Wide Web offered users a virtual home base online. It later became known as Yahoo!. Users who signed up at Yahoo! received an email address, access to a search engine, and news stories on their home page.

Meanwhile, a new company called Google launched in 1998. Like Yahoo!, Google offered a search engine. But it used a different algorithm to determine the importance of individual pages on the World Wide Web. Search engines made it possible for people to quickly find whatever content they wanted on the internet.

THE WEB TAKES OVER THE WORLD

Websites, along with all the text, images, audio, and video they contain, are known as web content. Writers, musicians, web designers, photographers, and videographers all create this kind of content. Many came to be known as content creators. As the web took off, people could find and consume web content faster, and they kept wanting more. This drew people's attention away from more traditional kinds of media. Journalists at newspapers and magazines still published printed materials, but they also began uploading stories to the web. Soon, web content took over as the leading outlet for media and information.

As people spent more time online, a new way of socializing emerged in the late 1990s. People used public chat rooms before moving conversations to private messages, phone calls, or email. AOL's "You've Got Mail"—a sound bite that repeated whenever a user received an email—entered the pop culture lexicon. Its instant messaging feature is the ancestor of today's DMs.

Smartphones, such as the popular Blackberry, could connect to the internet. But these phones were generally slow and provided a clunky experience. In 2007, a major invention changed that. Apple CEO Steve Jobs unveiled the iPhone. He described it as a "breakthrough Internet communications device."[3]

When working with his engineers, Jobs insisted that every action on an iPhone must take fewer than three steps.[4] By removing many of the barriers to smartphone

internet access, the iPhone brought more users online and transformed how people viewed web content. Instead of viewing web content through browsers, people began using online media applications.

Next came social media, a form of digital communication platform in which people create online communities to share ideas and content. Social media further changed how people consumed and interacted with information.

A few social media sites popped up in the late 1990s and early 2000s. These sites allowed users to create profiles and connect to other users. One early social media site was Myspace, which launched in 2003. The popular site boasted 25 million users at its peak in 2005.[5]

Social media sites geared toward all kinds of niches populated the World Wide Web. The blogging site LiveJournal allowed people to share their personal diaries online in real time. People shared photographs on Flickr, posted their resumes on LinkedIn, and watched homemade videos on YouTube.

Having the internet in your pocket with a real browser and real email, and the best implementation of Google Maps on the planet. . . . Having all this stuff in your pocket and yet having it be 10 times easier to use. I think this is where the world's going.[7]

—Steve Jobs on the iPhone, 2007

Steve Jobs introduced the iPhone at the Macworld trade show in San Francisco, California. He described the device as revolutionary and magical.

Eventually, Facebook overtook Myspace as the leading site for virtually hanging out with friends. In 2023, Facebook still reigned as the largest social networking site. It had more than three billion monthly users.[6]

The evolving technology of the internet continues to shape how people communicate. Social media adapts to each new development. For example, after the camera feature on iPhones made everyone a photographer, Instagram sprung up in 2010. It offered users a new way to edit and share photos online.

As dial-up shifted to broadband, data-sharing speeds increased. Broadband is a method of connecting to the internet that allows users access at higher speeds. People began to watch more videos. Sites and apps dedicated to sharing user-made videos, such as Vine, TikTok, and Snapchat, rose in popularity.

The internet has given more people the tools to share their ideas. Today, making web content is faster and easier than ever before. More people are making it and consuming it. With web content, people have the opportunity to be both creators and consumers.

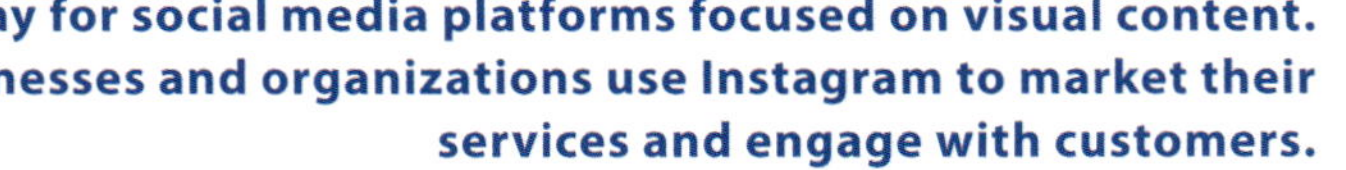

Instagram led the way for social media platforms focused on visual content. Today, many businesses and organizations use Instagram to market their services and engage with customers.

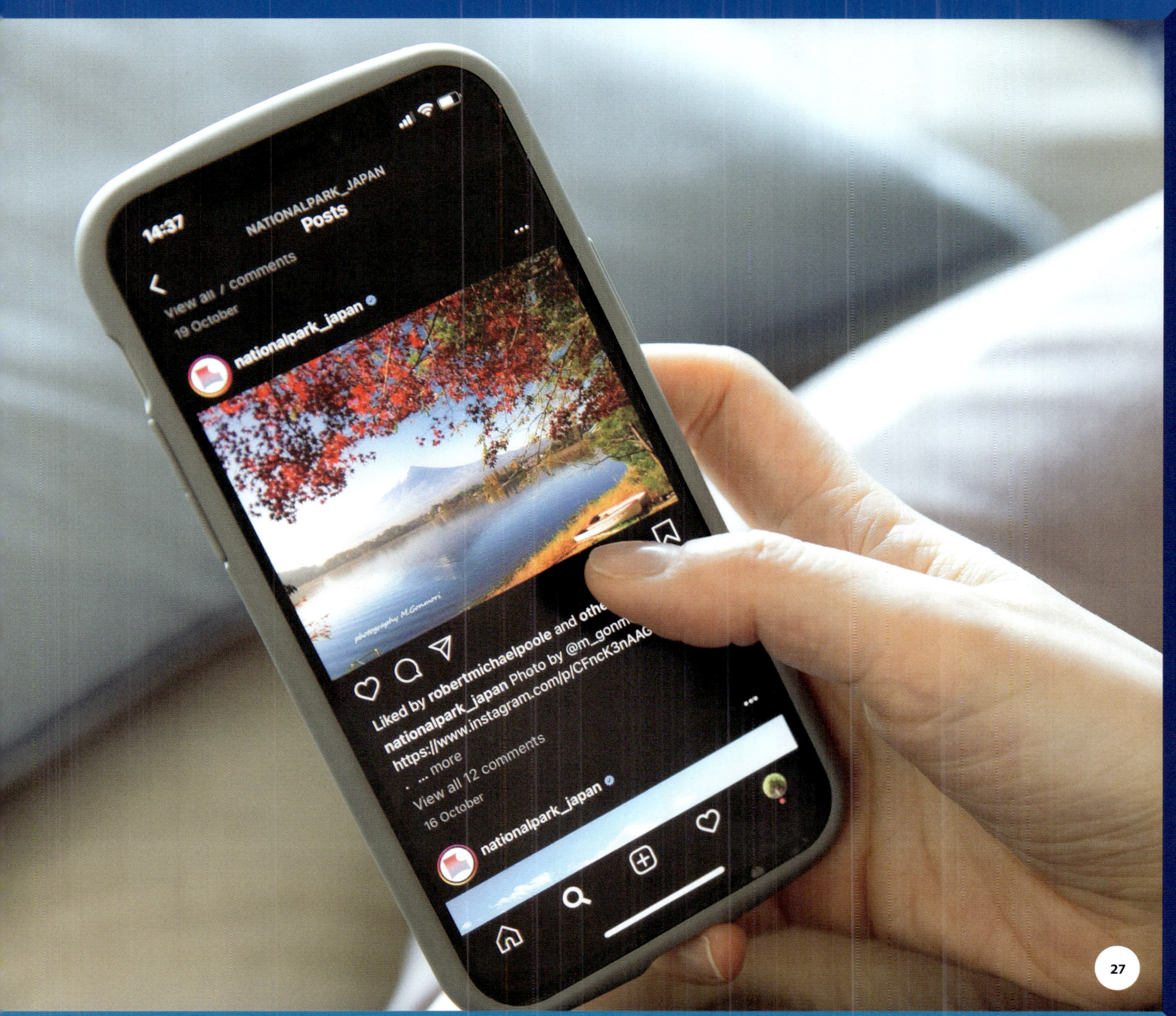

14:37
NATIONALPARK_JAPAN
Posts
View all 7 comments
19 October
nationalpark_japan
photography M.Gonami
Liked by robertmichaelpoole and othe
nationalpark_japan Photo by @m_gonm
https://www.instagram.com/p/CFncK3nAAC
... more
View all 12 comments
16 October
nationalpark_japan

BUILDING A WEB PRESENCE

Psychologists say people are drawn to create web content because the internet allows them to craft and display an outward persona. While people may not be able to control their real-life circumstances, they can have complete control over the profiles they present online. As more of the world's business and socialization goes digital, building a web presence has become a key part of creating a personal and professional public identity. Studies show that young people born after the internet became mainstream are most likely to express themselves through web content.

WHY PEOPLE MAKE WEBSITES

Some sources estimate that more than 250,000 new web pages are created every day.[1] Web pages are documents or pages on the web that can be accessed using a direct URL link. Websites are groupings of linked web pages accessed via domain addresses. Several factors drive people to keep creating new web pages

Many people build a web presence by creating content on multiple platforms. Disability activist Gem Hubbard has a website, a YouTube channel, and multiple social media accounts.

ACCESS TO THE WEB

It seems the only thing that has stopped people from creating and sharing web content is access. A 2005 study by Pew Research Center showed that teen content creators were more likely to have high-speed internet at home than teens who were not content creators. Teen dial-up users shared content at a much lower rate.[3] As technology advanced over the years, general access to web content expanded. In 2022, Pew Research found that 95 percent of teens surveyed had access to smartphones, while 90 percent had access to computers. Almost half of the teens surveyed said they used the internet constantly.[4]

and websites. People have always loved sharing information, which is the driving force behind the web's popularity. Studies show that sharing information and seeing how others respond helps people understand and process that information. What people share shows who they are and what they care about.

The internet allows people to create content faster than ever before. It is also easier and cheaper to create web content than it used to be. Plug-and-play platforms simplify the site-building process. They have made building a website intuitive, meaning people don't have to learn complex software.

Now that so many aspects of modern-day life are done online, personal websites often act as the faces people show to the world. According to Pew Research, about seven out of ten Americans use social media.[2] A profile page on a social media platform is similar to a personal website. The account owner creates all the content on their profile. Platforms such as Facebook and TikTok host the content and show it to others. But the account owner is responsible for any news, images, or text shared on the profile. People who create content that others like often build their own websites, especially if they have a career, business, or hobby related to the topic.

Building and maintaining a trustworthy website on a specific topic helps people establish themselves as experts on that topic.

BUILDING A PERSONAL WEBSITE

Web pages include more than words and pictures. Web pages can feature animations, audio, video, text, and games. Websites may be creative, functional, or both. There are many ways to make a website. Whether a person is setting up a complex business site or a small personal site, the process usually involves the same steps.

First, a person must identify the purpose of their content. This means figuring out the website's goal. The creator might consider what they want people to learn from their website. They must also identify their target audience. There is no point in putting a site online unless people are going to visit it.

Figuring out who will come to a site and what they will do there is important. It can help a creator decide the best structure and layout for their site. A site's target audience

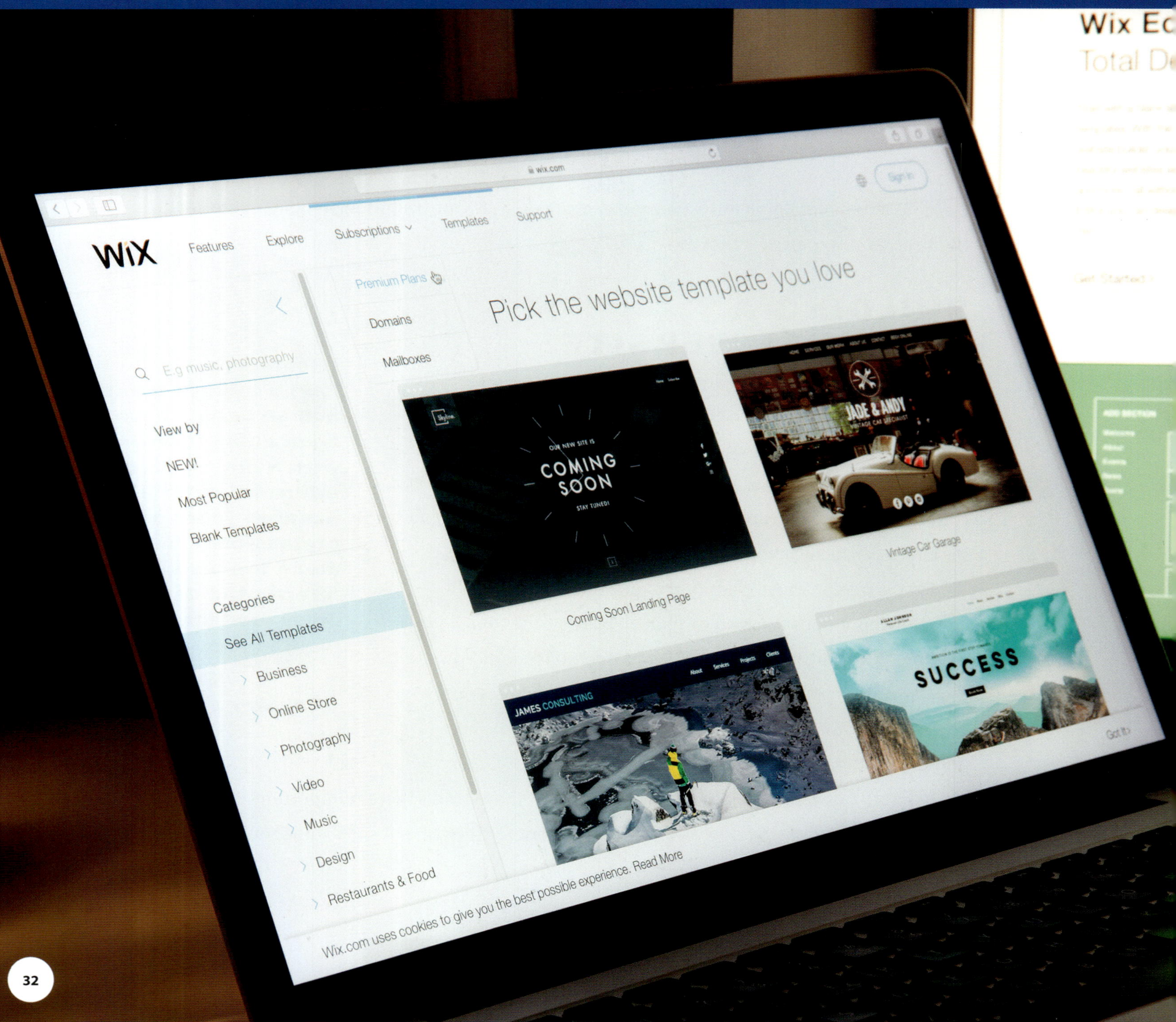
Wix Ed
Total De
wix.com
Sign In
WiX
Features
Explore
Subscriptions
Templates
Support
Premium Plans
Domains
Mailboxes
Pick the website template you love
E.g music, photography
View by
NEW!
Most Popular
Blank Templates
Categories
See All Templates
Business
Online Store
Photography
Video
Music
Design
Restaurants & Food
OUR NEW SITE IS
COMING
SOON
STAY TUNED!
Coming Soon Landing Page
JADE & ANDY
VINTAGE CAR SPECIALIST
Vintage Car Garage
JAMES CONSULTING
SUCCESS
Got It?
Wix.com uses cookies to give you the best possible experience. Read More

may include only a few people, such as hiring managers. But no matter the size of the audience, the content creator must consider it in their plans. One way for creators to figure out their target audience is by looking at similar sites and trying to gauge who the sites' audiences are.

The next step is researching and brainstorming. To start, a creator should make an outline or content plan. This involves sketching out ideas and building a blueprint of what will be on the site. This step is called wire framing because designers create the scaffolding they will use to build the site. Most websites include the same kind of pages, such as the "About," "Blog," and "Home" pages. But there may be some variation depending on the website's subject. YouTube creators such as Justin Kan have these pages on their YouTube profiles, since that's where their followers go.

However, having a website is important for people who are building a brand online. A platform such as YouTube can make changes that greatly affect creators, such as removing the monetization of certain content. Creators who have their own sites can still reach followers, while a person whose entire presence dwells on a social media platform is at the mercy of that platform's decisions.

After the planning stage, it's time to write the site's content. Creators should set aside separate time to write the text for their sites. Writing requires different skills than making designs or art. In addition to writing text, many people create multimedia content for

Many website-building platforms, such as Wix, offer hundreds of customizable templates. People can use these to set up their websites.

their websites. To make some types of content, creators may need to purchase special equipment or software. It may take much longer to create multimedia content if the writing hasn't been done first.

Next comes the editing stage. Website creators must edit for user experience (UX). This means making sure the site is easy for people to access, use, and navigate. The goal of web designers is to improve a site's UX. To do this, creators can explore the site as if they were a visitor. Using the site like a visitor may reveal blind spots in the design process. The creator must also edit the website's text. Doing a final proofread can help a creator catch any typos or mistakes before publishing the site.

LAUNCHING AND MAINTAINING A WEBSITE

After building a website, a creator must make sure it functions properly. Some images may appear in the wrong places or show up larger than anticipated. Some content may be cut off the screen when viewed on a cell phone. It's important that the content looks appealing on all types of screens and devices. Although most websites are built on computers, web content is often accessed on mobile devices.

While the testing phase might slow the process of making a website public, it is an important step. Search engines usually rank well-built, fast-loading sites higher in

WENDY JIE HUANG

YouTuber Wendy Jie Huang, who goes by the online name Wengie, always loved makeup and fashion—two subjects no one else in her family knew much about. So she tried to learn more on her own. She watched makeup tutorials but sometimes struggled to make beauty tips work for her because she looked different from most models in the tutorials.

Huang eventually started a job in digital marketing, helping clients start business websites and social media pages. She launched her own blog to get a better understanding of their experience. In 2013, she started her own YouTube channel. On her channel and her website, Wengie, Huang shares her views on beauty and style. One of her videos, in which she compared Asian and Western makeup styles, went viral.

Huang tweaked her content as she learned what people most wanted to see. She increased her following and leveled up with more support and better equipment. For example, once she gained enough followers, she stopped using a camera tripod and hired a videographer instead. Huang has released her own product lines, lent her voice to a Powerpuff Girl, and released original pop music. In 2023, Huang had more than 13 million YouTube subscribers.[6]

In 2019, Huang spoke at the RISE technology conference in Hong Kong.

BEWARE OF TROLLS

Sharing content on the web can be difficult. Content attracts attention. This attention may be positive, but it's not uncommon for content creators to receive negative comments. Some people post insensitive comments just to be a nuisance. These people are known as trolls. Content creators who post online should be prepared to encounter trolls. Trolls feed on reactions, so it's best to not respond to their comments. Creators should expect anonymous trolls and ignore them when they show up. If a troll's behavior escalates, the creator can block them and report their actions to the platform administrators.

search results. Testing a site ensures that it looks professional and functions smoothly, which can help drive more web traffic. Once a website is ready to share with the world, the creator can finally click Publish. This makes the site become live on the internet for people to see.

After a website is published, the creator must promote it. Without promotion, the target audience might not find the site. Creators can choose several promotion methods. One is search engine optimization (SEO). This involves using keywords within the site's body text and headers that show what the site is about. Keywords are common words or phrases people use to look up topics, so SEO pushes the website up in a search engine results list.

Stoney deGeyter works with SEO at a marketing company. "We tend to think of 'pleasing the visitor' as something that happens after we get the traffic, but good SEO puts this at the forefront of their efforts," he explained. "Search engines want to show only the best and most relevant websites to their searchers. Which means that instead of trying to please the algorithms, we should do what search engines do: try to please the searcher."[7]

Website creators can use online guides and tutorials to learn more about SEO. They can also do keyword research by typing words or phrases into search engines such as Google.

Some creators promote their sites by starting a mailing list or newsletter. They collect the email addresses of people who enjoy their site to let them know when they post something new. Another option is a blog. Most pages on a website can be static, meaning the information doesn't change often. But blogs must be updated frequently. Websites that are updated frequently rank higher in search results, so blog posts can serve as advertising for a site. Creators can also post links to their site on social media platforms and encourage people to share them.

Website creators must pay attention to engagement on their sites. If someone reaches out to a website, the website should respond. Website creators must consider how much visitors interact with the site. Engagement can be measured by how long visitors view a

Some content creators, such as beauty bloggers, partner with makeup brands. They feature the brand's products in their posts or videos.

piece of web content, along with clicks, downloads, and shares. Some website-building sites track this information for creators.

Creating a website is only the first step. Maintaining a site is just as important to its success. Web content attracts visitors to web pages, and updating the content keeps them coming back. Websites compete for viewers' attention, and the most compelling web content wins the most views. Creators must provide interesting content in order to attract views. The content must also be useful.

Search engines use algorithms to determine which sites go at the top of their lists. They rank search results based on what a user types into the search bar. For example, when a person searches "cars for sale," the search engine scours an index of content from millions of pages. It looks for web pages that contain information about cars for sale.

MAKING MONEY WITH WEB CONTENT

Some people make money by creating web content, even if it isn't their full-time job. The internet offers all kinds of avenues for aspiring creators. People who enjoy photography can sell their images to stock photo sites, earning a few cents every time someone downloads their photos. Designers can build apps that people buy and use on their phones and tablets. Designers can sell these apps in the same app stores that corporate creators use.

Creators may also earn money by partnering with businesses or companies. Sometimes businesses send out free samples of products to people who have a certain amount of followers on social media platforms. In exchange for the free items, the person promises to post videos, pictures, or comments about the items. Businesses may also offer a free item in exchange for a person's review of the product.

In the past, content often cost a lot of money to make. A large audience was required to justify this expense. For this reason, many pieces of content geared toward smaller audiences were never created. Today, modern technology makes it much easier and cheaper for people to make content. Most people need only a computer with an internet connection to get started. A content creator can earn a living creating content that is specifically targeted to smaller audiences.

This development has resulted in a vast array of content and creators to choose from. For example, some fashion content creators choose to focus on a niche, such as vintage fashion. Shirin Altsohn, Rachel Maksy, and other creators of fashion-related content share vintage-inspired outfits and hairstyles.

Content creators who make podcasts may need special equipment, such as microphones, mic stands, pop filters, and headphones.

Publishers Clearing House
WIN $7,000.00
a week
FOR LIFE!
I Want To Win!

BUSINESS ON THE WEB

In many ways, professional and personal web content are similar. Just like socializing and consuming media, business has largely moved online. A company's website is the home base of its web presence, serving as a digital reflection of its brand.

A company's web content is also a large part of its advertising. Today, many companies do much of their business on the web, especially aspects such as marketing and sales. Creating content has also become a business of its own, reflecting how much content people need and the things they need that content to accomplish.

WHY BUSINESSES NEED WEBSITES

Businesses build websites for the same reasons individuals do—to present a specific image of themselves to the world. Potential customers often judge a company's quality by its site's performance and appearance. An attractive and well-functioning

Some companies hire web designers to create their websites. Web designers may work on setting up the website's layout, creating graphics, and making sure the site is easy to navigate.

website adds legitimacy to a business. Professional websites can also make doing business easier, cheaper, and faster.

Most business sites follow the same basic structure, but they present information in their own unique styles. Most importantly, a business's website clearly explains what the business does and who its customers are. Professional sites often include several important pages. The "Home" page is what people see when they first visit the site. The "About Us" page provides a description of the company's history and mission.

Many business websites also include contact information, such as the business's phone number, email address, and physical address. They include links to the company's social media profiles too. Businesses may also create specialized pages that fit their focus or services. Bookseller Barnes & Noble, for example, includes specific pages for eBooks, toys and games, fiction titles, and bestsellers.

Another common page on a professional website is the shop page. Visitors can scroll through products, make purchases, or schedule services. It's common to see "Frequently Asked Questions" (FAQ) pages on company websites too. These pages provide answers to the site owner's most common queries. Some websites include a page dedicated to testimonials or press coverage. Testimonials are reviews made by real customers. A testimonials page might include links to published articles or videos that reflect the company's accomplishments.

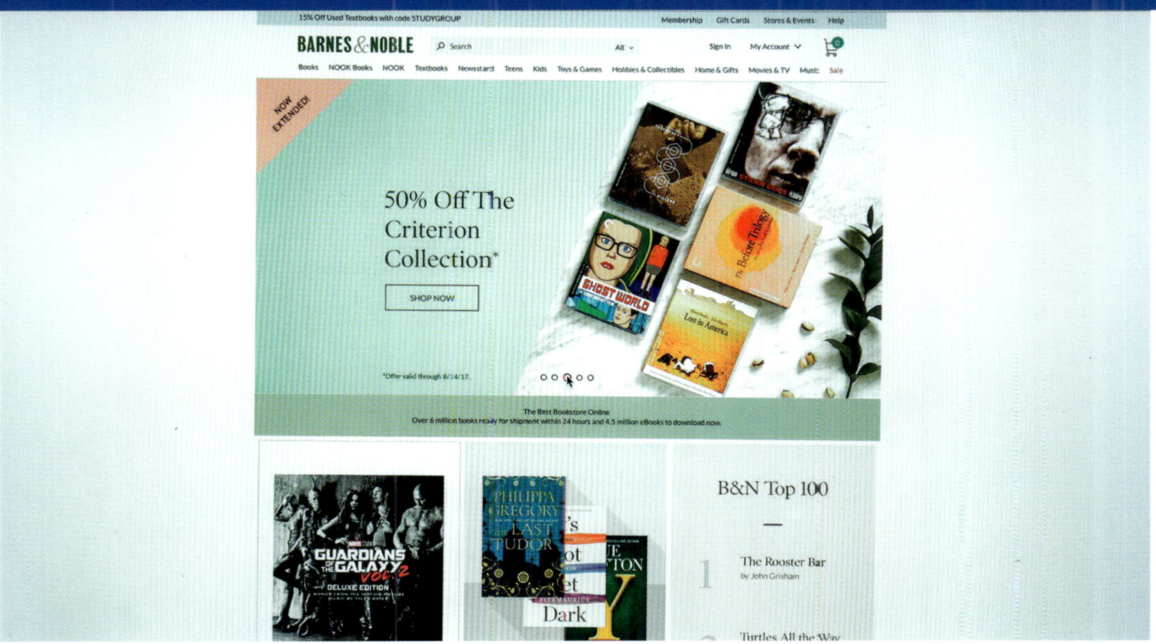

 **Barnes & Noble's website features pages that are frequently updated to reflect trends in the book world. Users can browse through bestsellers on the B&N Top 100 list, read articles on the B&N Reads blog, and check out the "Coming Soon" page.**

Websites have significantly changed how people shop. In some ways, websites have made shopping easier. Sites dedicated to selling things often use sorting functions to help consumers find specific products. People can sort or filter items by size, color, style, or brand. This allows consumers to find the items they're looking for and compare prices without having to make phone calls or visit stores in person.

However, online shopping has proven to be overwhelming for many consumers. Some think online shopping requires too many choices. In the information age, many people report a new problem called choice overload, in which the brain simply tires

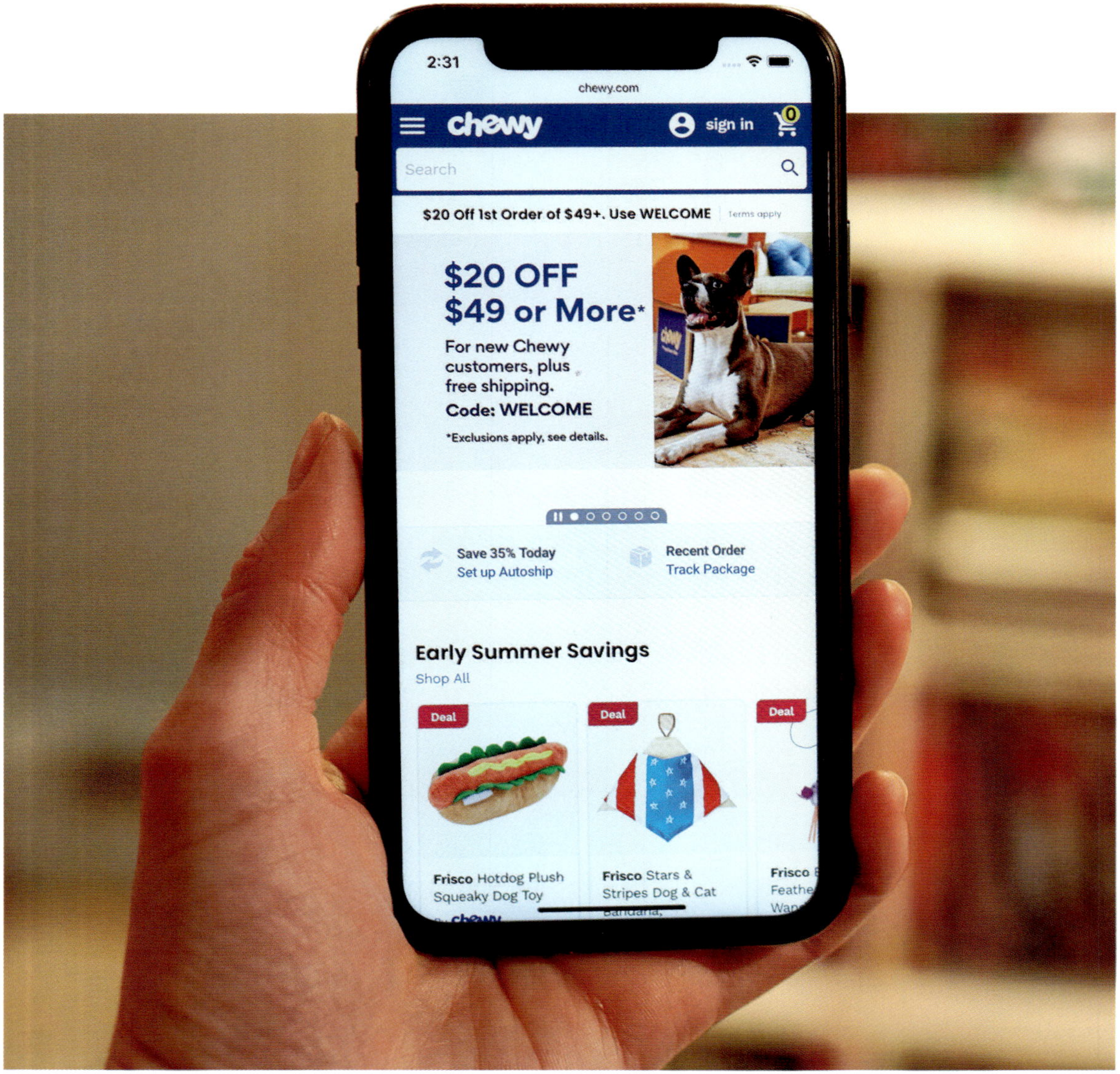

Many online retailers appeal to customers by providing good customer service and offering discounts. Pet product company Chewy, for example, sends personalized notes with shipments and offers brand-name products at reasonable prices.

of making so many small decisions. The instant gratification provided by smartphones has also resulted in increased demands from consumers. They expect an easy, quick online shopping experience. They also expect prompt responses from businesses they contact.

THE PURPOSE OF BUSINESS WEB CONTENT

A business's web content serves many purposes. While individuals use social media to socialize with others and consume media, businesses use their websites to strategically communicate with customers. The key difference in their web content is strategy.

Businesses create web content with a specific goal in mind: to get the user to do something. For example, a business might want its website visitors to buy a product or sign up for a mailing list. Whatever the advertiser wants done is known as a call to action. The quality of the content is evaluated by how well it converts, or gets the reader to answer the call to action correctly. Web content performs well when it successfully inspires users to answer a call to action.

SMASH THAT SUBSCRIBE BUTTON

Subscription models are on the rise as an e-business model. Owners of large databases of useful digital content may sell temporary access to their information. Content creators can also embrace this model by offering tiered premium content to fans. With more than 200 million subscribers, Amazon Prime normalized the subscription model. Small-scale creators can use the same principle to earn predictable, recurring income. For example, a podcaster may offer subscribers special access to ad-free episodes or bonus content.

Most businesses want their web content to do several things. Just like traditional advertising, web content should spread awareness and build demand for a product. Web content should also convince consumers to know, like, and trust a brand. Loyal customers are extremely valuable to a company, which is why businesses work so hard to appease upset customers. No business wants customers to spread negative feedback about their service.

Some professional websites publish content that helps their customers solve problems, such as how-to articles related to their products or industry. For example, a lawyer might post articles explaining in simpler terms how a law works and how it affects the average person. This content can help establish the lawyer as an expert on the subject. Every piece of content on a business web page helps support the business's mission. Web content is a modern door-to-door salesperson, reminding people how useful a product is and offering them a convenient way to buy it.

One of the main jobs of business content is to improve its site's rankings in search engines. People tend to trust the results that search engines provide. Almost half of the people using search engines visit the sites that show up at the top of Google's results. To put this into perspective, just slightly more than 10 percent of people bother to visit the second site in the list.[2] Once marketing departments learned the importance of getting to the top of the search results list, it became their main objective. Businesses can pay to be posted at the top of the list. This is called search engine marketing or paid search. While these postings look like other search results, they are often labeled as sponsored advertisements.

Sites don't have to pay to reach the top of the list, but many have figured out a way to create content that makes them more likely to get to the top. Each search engine uses its own algorithms to analyze a site's worth. Google, for instance, often prioritizes updates, or the last time new information was added to a site. With these competing factors, creating good web content is more difficult than it looks.

Making web content requires many technical and creative skills. Content creators must excel at making whatever type of content they specialize in. This may include video, audio, graphics, or text.

Bloggers need writing and storytelling skills, while vloggers need strong public speaking skills. Both need the technical skills required to run a website. They must be able to understand what content people will find interesting, engaging, and fun, as well as how to adjust their content based on audience reactions. They also require an eye for design and the determination to overcome obstacles during the creative process.

It's helpful to have experience in videography or photography, along with video or sound editing.

Since the process of creating engaging content involves so many skills, businesses tend to hire professional content creators. Some hire content creation teams. Other times, businesses outsource content creation to companies that specialize in creating business content.

CREATING CONTENT IS A FULL-TIME JOB

Hiring a professional can be expensive. As a result, many small businesses and sole proprietors create their own sites. Plenty of site-building platforms, such as Wix, walk people through the process. But some small businesses hire professionals or have a department dedicated to web content creation. They strive to make sites creative and fun for visitors. Professional content creators know how to balance creativity and function.

Professional content has a job to do. But it's important to remember that making money should not be a website's only objective. Some popular sites started out as personal sites. For example, film buff Col Needham started the website IMDb as a diary of all the details about the movies he watched. He stored the information in a private database on his PC.

In 1998, IMDb creator Col Needham partnered with Amazon. As CEO of IMDb, Needham has helped the website grow, expanding its offerings and representing the site at film events.

IMDb
IMDb

SPOTIFY

The music streaming platform Spotify is a good example of a professional business that tailors content to its customers. The site builds weekly playlists for users and recommends songs and artists based on what users have listened to before. At the end of each year, it also creates a customized Spotify Wrapped profile for users. This is an overview of all the songs, albums, and artists a user has listened to during the year. It also includes stats about the user's listening habits, such as their most-played song or album. Users can share their Spotify Wrapped profiles on social media. This serves as advertising for the Spotify platform.

Once the database reached 10,000 movies, Needham moved it to the web.[4] But he didn't earn money from the site. The site was built to compile and share information about a beloved hobby, but it garnered so much attention that it became a profitable asset.

Needham used his own funds to help maintain and improve the website. He used ad partnerships with movie studios to generate revenue. Needham inadvertently started one of the world's most popular websites by simply focusing on something he was passionate about.

HOW PROFESSIONALS PROMOTE CONTENT

In addition to creating content, businesses must also promote it. Companies are constantly looking for innovative ways to get their content in front of the eyes of potential customers. They can spread online marketing content in a number of ways.

Organic content often gets posted on the internet or on social media platforms just for fun. Sometimes this type of content spreads on its own when people share it. Sponsored content, on the other hand, refers to web content that a business pays other people to create.

Traditional web advertising can take many forms. It might appear as a banner ad at the top of a web page, advertising an upcoming concert. If someone clicks on the ad, they are taken to a site where they can buy tickets for the concert.

Another method is called native advertising. It involves creating ads that look like they are part of a page's regular content. For example, a business might create an ad that looks like content on a popular blogger's page. This entices the blogger's followers to click the ad with the same enthusiasm they have for the blogger's usual content. The blogger's fans are the business's target market, so it benefits them to reach this audience.

This practice has existed since long before the internet. Magazines have long featured advertorials, or sponsored content ads that look like regular articles. Today, people must follow guidelines in order to make these ads less manipulative.

BEHIND THE SCENES

Many people love to see behind-the-scenes content. A business can make simple videos inside their offices or warehouses to give consumers a better idea of its quality and internal practices. Consumers want to know that businesses treat their employees equitably and support important causes. A great behind-the-scenes video might feature interviews with employees or a thorough explanation of how a product is manufactured. The meal-kit company HelloFresh, for example, shares behind-the-scenes content on TikTok. Its videos have included employee profiles and a look at how food stylists set up photo shoots. These videos give viewers useful information about the business's inner workings.

Native advertisements must be clearly labeled to let readers know that they are paid for. The content requires a disclaimer explaining that the content is sponsored. Native advertising works well because it's less disruptive than traditional advertising. It attempts to slip into the viewer's entertaining content rather than distract them from it.

Some companies promote their web content through influencer marketing. This involves partnering with an influencer, or a person who is popular and well-respected among their target audience. A company may compensate an influencer for mentioning how much they love the company's product.

Content creators with big followings may also partner with companies. The business pays the creator to create content. Both the business and content creator share and promote the content. These arrangements are mutually beneficial. The content creator receives funding for expensive projects, while the business receives some promotion every time a viewer likes and shares the content.

Businesses may also use the internet to collect information about their customers, which is why they often ask visitors to share their email addresses. Email has become one of the primary ways of sharing web content with potential customers. Collecting information about customers helps businesses connect people to the items they want. Gathering email addresses allows a company to send targeted ads to a specific kind of customer.

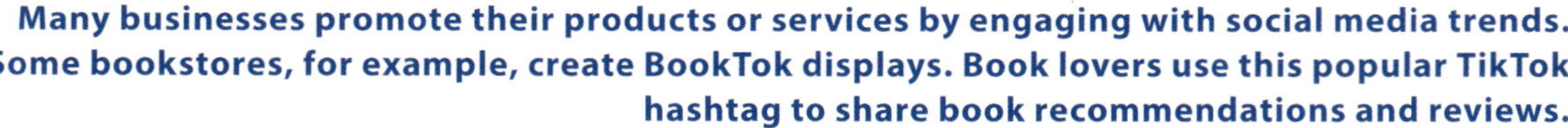

Many businesses promote their products or services by engaging with social media trends. Some bookstores, for example, create BookTok displays. Book lovers use this popular TikTok hashtag to share book recommendations and reviews.

ON SALE
SEPTEMBER 27
#BookTok
#BookTok
7:30
SHARON CAMERON
HIDDEN
PLACES
MARIE LU
DANCE
OF
THIEVES
MARY E. PEARSON
COLLEEN HOOVER
THE NEW YORK TIMES BESTSELLER
HOUSE
No. 1
THE
INVISIBLE
WAR
KISS
STEPHANIE PERKINS
THE
CITY
OF
BRASS

CREATING VIDEO CONTENT

Moving images and sound are more compelling to human brains than text and static images. It's also easier for most people to learn and remember information conveyed through video. Video content attracts such a large viewership because it's often more engaging than other types of web content.

Ever since high-speed broadband internet became available, video has been one of the most popular types of web content. iPhones put more than the internet in people's pockets. Today, almost anyone can record a video at a moment's notice. They can make all kinds of videos, from full-length YouTube videos to short TikTok clips.

A 2023 study looked at the habits of internet users. It found that about 98 percent of internet users between the ages of 18 and 24 regularly watch videos online.[1] The biggest online video platform is YouTube. Almost one-third of all internet users have visited the site.[2]

Some content creators post videos on multiple platforms in order to reach a wide audience. They might film long-form videos to share on YouTube and short-form videos for TikTok or Instagram.

TOOLS OF THE VIDEO TRADE

Most people own a smartphone. Anyone with a smartphone and an internet connection can make video content. Short-form videos are easy to create, post, and watch. A creator can simply grab their phone and film a scene in one take. Then they can download an app to a popular platform such as YouTube and upload the video. People have grown accustomed to amateur video quality, so extensive planning and editing isn't always necessary for making video content.

However, for those interested in producing higher-quality video content, upgraded equipment is essential. Creators must consider lighting, for example. Ring lights are an inexpensive way to achieve a more professional look in videos. These lights create a soft, diffused light that cuts down on shadows. Manufacturers now make ring lights that clip onto smartphones.

Modern smartphones have sophisticated cameras. But other types of cameras offer features that can't be found on a phone. For example, some videographers prefer to use a digital single-lens reflex (DSLR) camera. It allows users to switch out different lenses. They can also use the auto-focus feature or control options such as shutter speed. DSLR cameras use a mirror to direct light into an image.

Some modern mirrorless cameras are similar to DSLR cameras but use fewer components and offer more features. They are usually smaller and quieter too. DSLR cameras have a longer-lasting battery life and more optional accessories and lenses. Mirrorless cameras have more stabilizing technology, meaning they create more stable videos.

But videographers can also use a tool called a stabilizer to hold their cameras still. People can buy stabilizers made to use with smartphones. Improvements in smartphone video quality can make it difficult to tell the difference between videos shot on smartphones and videos shot on professional cameras.

SHORT-FORM VIDEOS

Short-form videos are effective for several reasons. Most people prefer videos to be less than a minute long, especially when visiting social media platforms.[3] People on social media are usually not there to watch long videos. Instead, they scroll through looking for shorter content.

Short-form videos, or microvideos, have risen in popularity because more people watch video content on phones and tablets. Short videos can make a point in just three to

30 seconds, so several videos can be easily consumed in little time.[4] People tend to hit up social media during quick breaks, so it's easier to reach this kind of audience with shorter videos. Social media apps such as TikTok and Snapchat are designed for short videos and have made these videos even more common.

Many content creators prefer making short videos because people consume them quickly, and there is always demand for them. However, short videos have some downsides. Short-form videos are less profitable than long-form videos and are generally considered to be lower quality.

Furthermore, some critics worry that the popularity of shorter videos is making the human attention span shorter. Microsoft Canada studied how technology affects attention spans. It found that from 2000 to 2013, the average attention span of young adults in Canada shortened from 12 seconds to eight seconds.[5] But for beginners, short-form content may be the easiest place to start.

LONG-FORM VIDEOS

Long-form videos still have a place in the world of web content. People didn't always have the data speeds to make long-form videos available to the masses. But technology caught up, allowing videos to be longer again. A fifteen-minute video and an hour-long video would both be considered long-form. The most successful long-form videos hit

TikTok video trends come and go quickly. Popular trends on the platform have included dances and lip sync challenges. Many content creators participate in these trends to boost engagement.

two important factors: they are engaging and packed with useful information. Today, several popular genres of long-form videos are available online.

Let's Play videos are one common type of long-form video. These involve people playing video games while adding their own real-time commentary. The live streaming platform Twitch, which launched in 2011, is known for featuring Let's Play videos.

Generally, the gameplay in Let's Play videos takes up most of the screen, while the vlogger may be shown in the corner of the screen. As the vlogger plays, they react to and comment on things happening in the game. While this type of video relies on someone else's gaming content, the vlogger's personality, gaming skills, and comedic commentary are often the main attraction for viewers.

Many gaming consoles now come with preinstalled software that can record gameplay. But creating professional-looking Let's Play videos requires expensive equipment such as microphones and cameras. Gamers can use external devices called capture cards to record their gameplay. Then they can prepare it for the web using film-editing software such as iMovie.

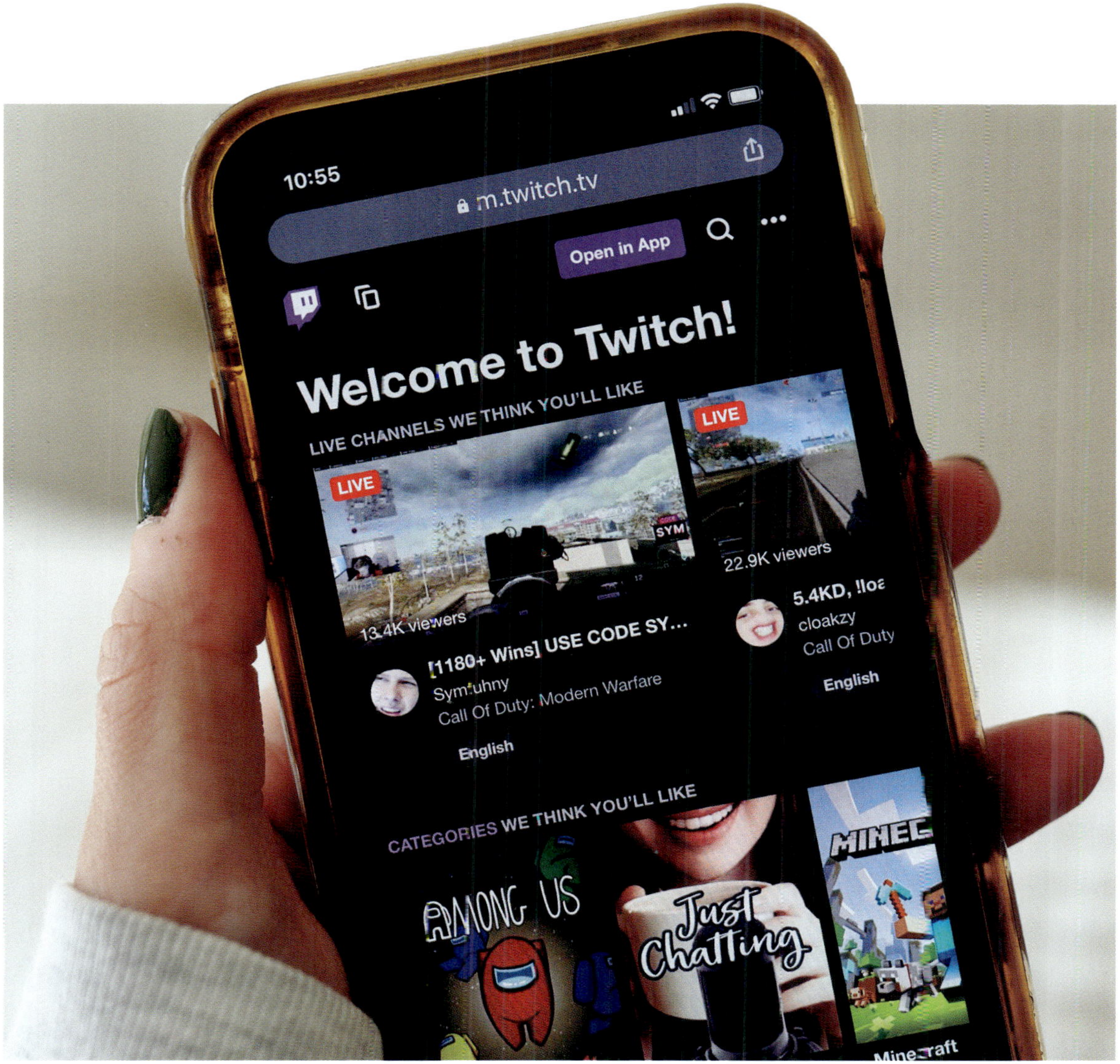

 In 2023, the live streaming platform Twitch had 140 million monthly active users. More than seven million people live streamed on the platform.

Unboxing videos are another popular kind of long-form video. These show a person opening a product or package for the first time, describing each item in detail while showing it to the viewer. Creators who make unboxing videos recreate the uplifting bliss that some people experience while shopping. Good unboxing videos capture the thrill of opening a gift.

While paid advertisements show only what a company's marketing team wants the public to know, unboxing videos describe the good, the bad, and the ugly. After the unboxing, the vlogger often demonstrates or uses the product for the first time, evaluating the pros and cons so that viewers can make better-informed buying decisions. Other long-form video genres include how-to videos, vlogs, video podcasts, and webinars. Innovative creators will likely develop even more types.

You design a ritual of unpacking to make the product feel special. Packaging can be theatre, it can create a story.[7]

—Walter Isaacson, on the packaging of Apple products

Users can also find more serious, high-quality video content on YouTube. This type of content is produced much like a TV show. For instance, Dave Amos is an urban planner and assistant professor at California Polytechnic State University. He runs the YouTube channel City Beautiful. It focuses on topics such as urbanism, design, and transportation. In 2023, the channel had more than 650,000 subscribers.[6] To produce each video, Amos needs to write a

MARQUES BROWNLEE

In 2009, teen Marques Brownlee wanted to purchase a new laptop. He watched several review videos before deciding which one to buy. After purchasing one, Brownlee noticed a few laptop features that no one else had talked about. So he made his own YouTube video to discuss them.

Brownlee called his channel MKBHD, which represents his name plus the initials for "high definition." He enjoyed helping people weigh their options before purchasing products. So Brownlee kept posting review videos of technology products and software.

Over time, Brownlee hired staff to help create his videos so he could make more of them. In 2023, the MKBHD channel had about 18 million subscribers.[8] Named Creator of the Decade at the 2018 Shorty Awards, Brownlee manages a team of cinematographers, writers, set designers, and other experts who help him bring his ideas to life. While he still reviews state-of-the-art technology products, he has also interviewed major figures in the technology world, including Bill Gates and Elon Musk.

In 2023, Brownlee spoke at the Samsung Galaxy Creator Collective event. He discussed his successful YouTube channel and how he got started making content.

Running Tunes

script, capture footage from various sources, build graphics, narrate the video, edit it, and then publish it to YouTube. He posts new episodes regularly.

VIDEO CONTENT STRATEGY

Getting started with video content creation requires a few basic skills and equipment. Creators must have a good camera, along with accessories such as camera tripods or microphones. They may also need to set up a place to record their videos. Video creators should be good at talking to the camera and explaining information in an understandable, engaging way. To stay on topic, it can be helpful to write a script. It's also important to have some knowledge of video editing. Creators can use editing software to cut, trim, and add special effects to their video footage. This helps make a video look polished and professional. Adobe Premiere Pro and Final Cut Pro are two common video editors.

Today, aspiring videographers can post and share their videos on a number of platforms. Beginners usually start by posting videos on social media for their friends and family. Creators who want to start posting more publicly should pick whichever platform best fits their content.

YouTube is the most-used video platform on the planet. According to its company mission statement, it strives "to give everyone a voice and show them the world."[9]

Content creators using Final Cut Pro can cut, trim, and reorder video clips. They can also add special effects, transitions, and music to their videos.

When YouTube was first made, it removed many barriers to sharing and finding videos on the web. It's the second-most popular search engine after Google, meaning more people use YouTube to search for web content than sites such as Bing and Yahoo!.[10]

When it comes to finding video content online, most people think of YouTube. But there are other video-sharing sites too. One is TikTok, which has more than one billion active users.[11] It started as a site focused on short music videos. The site's viewers tend to be younger than viewers of other video-sharing sites. Another site is Twitch, which is popular for Let's Play videos and streams. This platform allows users to broadcast their gameplay while interacting with like-minded creators.

When it comes to making web videos, practice makes perfect. Once amateur videographers start posting, they will likely want to expand their viewership. The next step is planning how to grow an audience. This process requires web strategy, or planning content with a purpose in mind. Most people set out to make videos for fun, which sometimes accidentally leads to success. But other creators decide what they want to achieve and then create a plan to make it happen.

To get started, video creators should do research by watching videos that are similar to what they want to make. They should pay attention to the audience of those videos, noting who they are and what types of content they share most often. It's also important for video creators to be consistent. They must set up a posting schedule and stick to it. For example, a YouTuber might post a new video every Friday. This practice builds trust with an audience. The key to building an audience is to remember that the internet is based on the principle of sharing.

Creators can also hang out in online communities they want to be a part of, sharing useful information and content. If a trend is going viral in a creator's industry or community, the creator should participate in it. Making a video aligned with a viral trend can expose a creator's content to millions of people who might not normally see it. The most successful amateur video content captures the audience's attention as fast as possible while feeling relatable. Beginners should try not to get too caught up in the business side of content creation. The key to creating any type of content is to enjoy the process.

PROFESSIONAL ONLINE VIDEOS

Many businesses choose to make video content because it provides the highest return on investment (ROI), compared with other kinds of web content such as text and images. ROI refers to how efficient or profitable an investment is. Marketers spend money on video content because it makes them the most money in sales. Organizations such as nonprofits may also use video content as a way to bring more attention to their missions or causes.

As creating video content becomes more feasible, more businesses and organizations are incorporating videos into their marketing plans. The growth in marketing video content is no mystery. Consumers report that watching a video about a product often helps them decide whether to buy it.

Many video creators have made a career out of creating web content too. Some creators even hire staff. They work with their teams to create high-quality videos. Then they post the videos on popular video-sharing platforms such as YouTube and TikTok.

Chef Sohla El-Waylly, *right*, appears in YouTube videos about cooking. In 2020, she had her own show on the YouTube channel Babish Culinary Universe.

STREAMING FROM SPACE

The National Aeronautics and Space Administration (NASA) has updated its web presence for the new generation of the internet. Visitors on NASA's website can view images from space updated daily through satellites. This includes information such as the location of active wildfires, temperature and precipitation rates, and the current quantity of ice in the North and South Poles. Anyone who wants to become a climate researcher or activist can now access NASA's technology to track the effects of climate change in real time.

THE DIFFERENCE IN GOING PRO

The main difference between an amateur and a professional content creator is strategy, or the content's goal. Creating the web content is not the end goal. The goal may be persuading people to sign up for a mailing list or buy a product.

Or the goal may be to make money, because the more people that watch a video, the more money a creator earns from the platform. Professionals test a video's performance to see whether the video succeeds in reaching this goal. Businesses analyze key performance indicators (KPIs), such as views, clicks, customer engagement, and sales, to decide if a video is successful.

Most professional video content is created by teams of experts. Each expert specializes in a specific aspect of video creation. Much of the equipment and skills needed to make web video content are accessible to individuals. But professionals know more about video showmanship.

Anyone can shoot a short, shaky video. But professionals consider how every aspect of a video's cinematography, or motion picture photography, represents a company's brand and message. This requires time for planning, equipment for making the video magic happen, and a team of people to bring it all together.

Many businesses have departments dedicated to creating marketing videos. Some hire professional videographers to develop their content marketing plans. In these situations, creators may present several ideas to a company. Then they work together to choose the best one.

Most professional video creators say the most time-consuming part of the process is preproduction. This stage includes brainstorming, writing, storyboarding, and casting. After this stage comes production, which involves recording or shooting the video. The last step is postproduction, during which the creator edits the video and prepares it for publication. Creating professional video content can be as complicated as producing a film.

"Data is really at the heart of everything we do. With YouTube, you've got an audience there that literally tells you whether they want to watch something or not, in real time."

—Richard Hickey, chief creative officer at Moonbug Entertainment

PROFESSIONAL EQUIPMENT

To create high-quality videos with complicated cinematography, professionals may invest in expensive equipment. They often use multiple cameras to cut to the most compelling

angle at any moment in a scene. While a large majority of amateur web videos are shot on smartphones, professionals are much more likely to use devices dedicated to capturing videos. About 70 percent of professional video content creators use DSLR cameras.[2]

While smartphone videos can be shot and edited on the same device, professional video editing requires software. Many professionals use Adobe Premiere Pro. With this software, they can generate video transcripts, add graphics such as the title and credits, and adjust color and light. Another popular video editor is iMovie.

Professional video-making may require special microphones too. Smartphones cannot always record high-quality audio, so professional videographers often use microphones to record audio separately. Then they sync it to the video footage. While this process is much more time-consuming, expensive, and complicated than shooting smartphone videos, it helps bring the magic of television and movies to web videos.

PROFESSIONAL CHANNELS

There are many professional video channels on the web. The TED Talk channel on YouTube shares the most well-received presentations at TED Conferences. TED is a nonprofit organization committed to finding "the most interesting people on Earth and letting them communicate their passion."[3] At the organization's TED Talk events, leading thinkers come together to share ideas about technology, entertainment,

Some video content creators set up home filming studios with professional equipment. This may include special lighting, microphones, headphones, and camera tripods.

and design. Each TED Talk packs a useful lesson into 18 minutes.[4] The gathering allows experts from different fields to share some of their recent findings with the audience. Giving a TED Talk has become a rite of passage and mark of success for many professionals. Highly shareable TED Talk videos can bring wider audiences to speakers who might struggle to gain attention. TED now offers its videos via its own apps.

Other professional channels are geared toward kids. The CoComelon channel streams children's entertainment, approaching videos for kids in a new way. The channel is owned by Moonbug Entertainment. This company searches platforms for existing children's video content and analyzes how videos perform. Then it purchases the videos with the most potential, investing to make them bigger and better. CoComelon proved so likable that Netflix purchased Moonbug Entertainment. CoComelon has become one of the network's most-watched shows.

Some professional content creators have made extremely successful careers out of making videos. James Stephen Donaldson, known on YouTube as MrBeast, rose to online fame by posting videos about gaming, internet drama, and fun stunts. After a video of

At the 2023 Nickelodeon Kids' Choice Awards, YouTuber MrBeast won the Favorite Male Creator award. He is one of the most-subscribed YouTubers in the world.

JUSTIN KAN

In 2007, Justin Kan began streaming his day-to-day life on his website, Justin.tv. He called this practice *lifecasting*. Kan's lifecasting was featured on major news outlets, which brought lots of attention to his website. His content jumped in popularity.

Soon other people wanted to stream videos of their lives, so Kan turned his site into a platform where users could share their videos. Soon, the Justin.tv site splintered into different sections. Each section featured videos focused on specific topics, such as entertainment, news, and gaming.

Gaming outperformed all the other sections, so Kan launched TwitchTV. This site featured only gaming content. Over time, it turned into the live streaming platform Twitch, which Kan eventually sold to Amazon for $970 million.[6]

Since then, Kan has launched a few other projects. But he mostly shares expert insights on new technology in videos on his YouTube channel. He also invests in the startups of young technology entrepreneurs whose ideas he believes in.

In addition to running his YouTube channel, Kan has started a podcast and founded several technology companies.

Donaldson counting to 100,000 went viral, he hired his friends to help develop and run new channels. Considered one of the most successful YouTube influencers, Donaldson produces his videos much like a production company makes a TV show.

He also puts popular keywords in his video titles. This is a common method of manipulating search results. Many of Donaldson's expensive stunts are funded by his viewers. He has used his success to fund humanitarian projects, such as removing 30 million pounds (13.6 million kg) of trash from the ocean.[7]

Professional video content creators earn a living from the content they create. They may partner with a brand by agreeing to use the brand's products or mention it in a video. Videos might also have affiliate earnings. This means that every time a viewer buys the featured product, the creator receives part of the sale. A creator can also make money by selling official merchandise in a shop on their site. The YouTube Partner Program asks video makers if they want ads shown with their videos. In exchange, the video makers receive a portion of the advertising revenue that YouTube earns.

E-COMMERCE

Selling and buying on the internet is called e-commerce. Shopify is an e-commerce platform that allows individuals to easily build online stores where they can sell goods and services. Whether someone wants to open an exclusively online shop or build a web presence for a physical store, Shopify helps them with a variety of tasks. This includes managing inventory, building a digital storefront, and facilitating electronic financial transactions. Shopify has been used to build more than four million e-commerce stores.[8]

INFLUENCERS

As social media made sharing content accessible to more people, a new kind of entrepreneur was born—the influencer. Influencers are popular content creators or social media stars who promote products or services to their online audiences. Some influencers have unique knowledge of the topics they post about. Many gain a huge online following because of their personalities and ability to connect with others. The idea of influencer marketing goes back a long time.

Throughout the 1900s, businesses frequently partnered with well-known icons such as movie stars. But today's influencers are a relatively modern phenomenon. More and more often, advertisers partner with influencers because their social media posts influence customers' buying decisions.

In a 2019 survey, 86 percent of young Americans reported that they wanted to be online influencers.[1] Influencers having fun has become the new prime-time entertainment. While influencers portray ideal lives in their content, they also bring

Many successful influencers, such as Emma Chamberlain, are popular because of their relatable personalities. Since starting her YouTube channel, Chamberlain has appeared on talk shows, partnered with multiple brands, and started a coffee company.

attention to merchandise relevant to their niche, or subject matter, just as movie stars have served as spokespeople for companies in advertisements.

DEFINING AN INFLUENCER

Influencers and content creators are different. An influencer is a person who has the power to influence the behavior and decisions of large groups of people. Influencers usually share the work of content creators on their social media platforms. Content creators actually produce the content. They shoot photographs, record videos, write blogs, and produce beats. Some people are both content creators and influencers.

Hootsuite is a company that helps users manage multiple social media accounts. It has found that the top influencers are internet personalities who post frequently and have more than one million followers, several well-known brand partnerships, and high engagement from followers in the form of likes, comments, shares, and views.[2] For example, influencer Huda Kattan started a beauty blog

HANNAH MELOCHE

Hannah Meloche started posting content when she was 13 after years of practicing on her mom's phone. She was recognized by a viewer for the first time at a football game during her freshman year of high school. Hearing positive feedback from people who enjoyed her videos motivated Meloche to keep posting. On Instagram, she posted photos and goofy videos of herself with her friends, whom she called "the girdies."

When videos surfaced of Meloche making fun of a fan, she owned up and apologized for her behavior. She learned a lot from the negativity that surrounded her. Despite her seemingly picture-perfect life, Meloche gets honest and vulnerable in one of her most talked-about videos. She tells viewers about the pressure she feels to appear perfect, saying, "It gets overwhelming and I think people forget that YouTubers are real people too."[4]

While she lost some followers, her authenticity resonated with others. Meloche now says the overall goal of her content is to bring more positivity to people's lives. Meloche also fulfilled one of her lifelong dreams by starting her own jewelry business, Starlite Village.

Meloche is known for posting travel, fashion, and lifestyle content. In 2023, she had about two million YouTube subscribers.

CANCEL CULTURE

The word *cancel* **refers to when people remove or encourage others to remove support for a public figure because of that person's words or actions. Anyone who puts themselves in the public eye can get canceled, which can mean losing followers' respect, admiration, and attention. Expressing support for a controversial view or defending a person who exhibited unacceptable behavior can get someone canceled. However, this does not always have concrete, long-term consequences. Critics say the rise of cancel culture punishes free speech. But it also reflects a growing trend in transparency between public figures and their audience.**

in 2010, which she built into a brand, Huda Beauty. To market the brand, Kattan posts videos of herself using Huda Beauty products. Khaby Lame got his start making funny TikTok videos in 2020. By 2023, he had partnered with fashion giant Hugo Boss, featured NBA stars in his videos, and become the most followed person on TikTok.

Good influencers often make it look like they are getting paid just to have fun. But a lot of work goes into making success look easy. An influencer is an expert who excels at developing authentic connections. Successful influencers manage to build a good reputation for having great insight and ideas about a specific topic. Followers stick with influencers because they trust them, and influencers build that trust by sharing useful information and being honest.

Many influencers share just enough about themselves to make them relatable. Good web content engages viewers, meaning it gets them to look at their devices and pay attention. Influencers are masters of knowing what kinds of content will engage their followers. They are also masters at making their followers feel seen. They answer comments, share posts about causes their followers care about, and may make changes and heartfelt apologies based on their followers' criticism. Influencers have likable personalities and the ability to attract people, but they might not be faking it. Internet audiences can sense a phony, so authenticity is one of the most important aspects of gaining the power of influence.

BECOMING AN INFLUENCER

Aspiring influencers can start by becoming an expert in a subject they love. This is important because sticking to a topic over a long period of time requires drive and excitement. Some in-demand topics are sports, gaming, makeup, fashion, and humor.

Creators should post content on social media platforms that they use for fun. They should pay attention to which types of content people share. The creator can also build a home base, or a personal website that includes links to their social media profiles and their contact information. Businesses can use this information to get in touch with the creator.

Influencers try to get their audience to not just watch or read their content but also engage with it. They often encourage people to share, like, or comment on the content. Influencing is all about connecting, so influencers should take the time to learn about

RATIS
alita
GRATIS

their followers and talk to them. They can view the content their followers post too. Influencers get to know their followers in order to get better at finding the types of content the followers love. As their audience grows, influencers become more appealing to brands that want to market to the audience.

When brand deals start rolling in, influencers must remember the importance of transparency. They must follow all the relevant laws and regulations concerning influencer marketing and advertising. For example, when a person is compensated to make content, the poster needs to disclose their involvement in a paid partnership.

HOW INFLUENCERS MAKE MONEY

Across the board, influencers earn vastly different levels of income. Most earn very little. Some influencers create content in exchange for free samples of products through companies such as BzzAgent. The biggest determining factors in an influencer's success are the size of their audience and the amount of

Fashion influencers often focus on posting high-quality photos of their outfits. They may use special cameras or photo editing tools.

engagement they receive. There are five main levels of influencer reach. Nano-influencers have fewer than 10,000 followers, while micro-influencers have around 10,000 to 50,000 followers. Mid-tier influencers have a following of 50,000 to 500,000 people. Macro-influencers have 500,000 to one million followers. The highest level of influencer is the mega-influencer. These influencers can have more than one million followers.[6]

Many influencers have multiple income streams. Income is any revenue or money that a person earns. Influencers earn money in a variety of ways. These different methods of earning are the streams that flow into the influencer's total earnings.

Influencers often have passive and active income streams. An active income stream involves performing work, such as jobs at a company or business that may be unrelated to their content creation. For passive income streams, less work is usually involved. As an example, a creator makes a video once and earns money through ad revenue every time a viewer watches the video.

An influencer's social media presence can sometimes lead to profitable opportunities. For example, a makeup brand may reach out to a beauty influencer about

Sometimes popular influencers are invited to events. In 2022, YouTube hosted a content creator event for Asian Pacific American Heritage Month.

a potential partnership. The influencer earns free samples of the brand's products if they feature it in their videos or posts. This kind of partnership may help increase the influencer's following, possibly leading them to start their own line of makeup products.

Influencers who build their audience on a single social media platform run the risk of losing followers if that platform changes its algorithms or terms of service. The influencer may also suffer a pay cut if the cost of getting views suddenly skyrockets. Influencers who build their own brand can still earn money by selling their own merchandise or posting content on another platform. Others may use a site such as Substack to sell subscriptions to premium content that is unavailable anywhere else.

Shorts
Dream Screen
AI-generated backgrounds in Shorts
Matthew Simari

THE FUTURE OF WEB CONTENT

The World Wide Web fundamentally changed how people create and consume content, and the transformation is still underway. The children who grew up in a wired world are now old enough to deconstruct the information superhighway and rebuild something better. What that will look like remains to be seen, but technology insiders have some predictions.

THE SHIFT TO WEB3

Computer scientist Gavin Wood created the term Web3 to define the future of the internet. Web3, or Web 3.0, refers to the third wave of the internet. The first generation of the internet, 1.0, revolved around creating static content on web pages that people could visit and read.

Web 2.0 brought the principles of engagement and interaction, where information on the internet flowed in both directions. Users could not only view pages but also interact with other people's information. This change spurred the

As the internet continues to evolve, many companies are experimenting with new technology. In 2023, YouTube announced Dream Screen, a new YouTube Shorts feature that uses artificial intelligence to generate background images.

growth of social media, which encouraged people to like, share, and create content by providing tools that made creating content easier. The platforms that made it easy to create content monetized that service by selling people's data to businesses that could use it to advertise more efficiently.

Web 3.0 will prioritize decentralization of data through blockchains. While the first computers centralized data on a single server that other computers could access, blockchains distribute data on several networked computers. No individual can change the data without the rest of the network's knowledge.

Blockchain technology allows for more transparency, since one person cannot change or delete data without being traced. It also provides a more secure way to store data. Today, people can hack into a server to access a company's data. But hacking into a blockchain would be much more complicated, if not impossible.

Web3 will also give people more control over their own data, including content. Web3 aims to remove some of the parties in the middle that censor, control, and profit from other people's content. The parameters of what can be shared will be determined by the community rather than a site owner.

Web3 could also fundamentally change global trade. Different countries may move away from their current financial systems and adopt cryptocurrency models, non-fungible tokens (NFTs), and other new forms of global currency. Developments in the way people pay for goods and services may open up new ways for fans to support their favorite content creators.

ARTIFICIAL INTELLIGENCE

A major factor in the future of web content is how artificial intelligence (AI) will affect how people create and consume content. Large language model (LLM) is a form of AI that can generate text in a humanlike manner. An LLM does this by analyzing information the engineers put in its database.

AI databases often include works the original creators did not consent to being analyzed or transformed by AI. Engineers are developing software that learns to be more accurate by repeating processes and receiving feedback on their output. Some critics argue that LLMs will lower the demand for organically generated content.

As AI evolves, it's hard to tell where technology like LLMs could take creators in the future. Some people think AI will make people rich while they sit back and do nothing. But it's more likely that people will continue to do most of the creating, and they may use AI tools to better target and edit messages for the creator's desired outcome.

HIGH-QUALITY CONTENT FOR ONE

Future web content will be tailored and personalized to the reader. AI will likely be able to use vast amounts of data on the web to perform research and explain it in a conversational tone. In the future, these technologies may prepare news feeds specifically curated for an individual's interests. Some people envision a network that could instantaneously create video content made to the viewer's requests. In other words, a person could tell their phone which movie they want to watch, and AI could instantly generate the exact video for the viewer to enjoy.

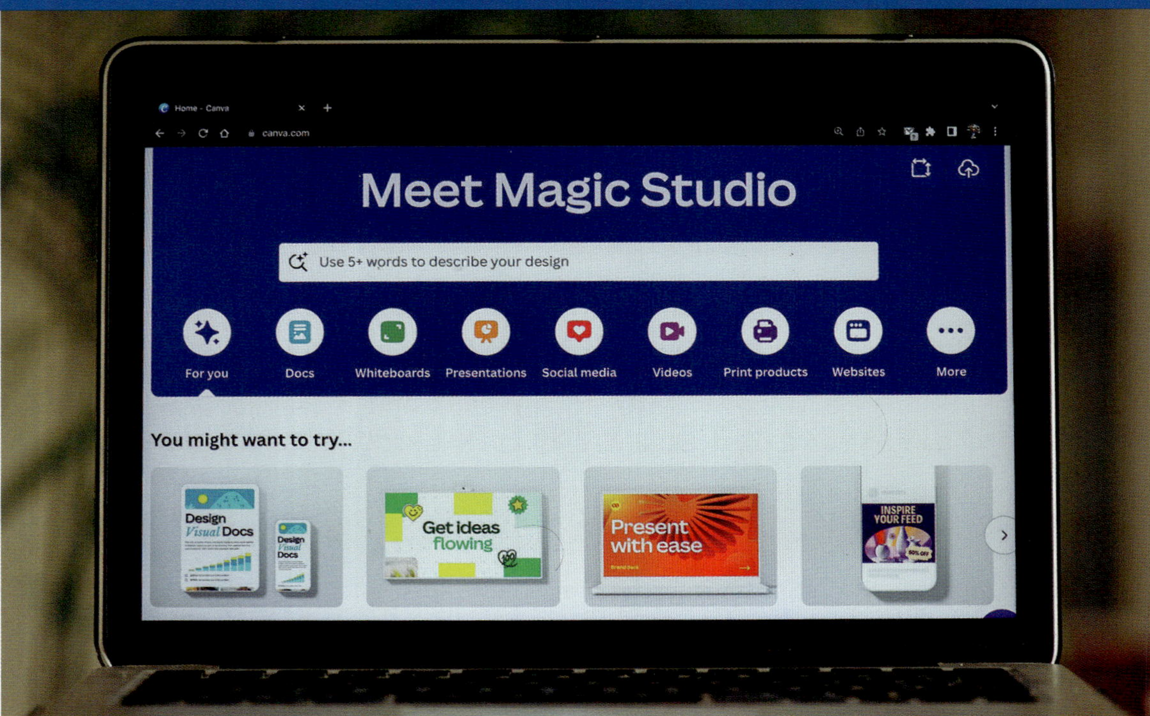

In 2023, the graphic design platform Canva released new AI tools. With the platform's Magic Studio, users can generate images by simply typing in what they want to see. 

AI's ability to analyze and suggest improvements based on prior performance and input may increase the quality of web content, but not without the help of humans. Creator Chris Bailey uses AI to help point out biases in his writing so he can improve his content. He posted a disclaimer on his blog to explain exactly how he will and will not use AI in his content creation. YouTube, Photoshop, Canva, and other types of content creation software have incorporated AI tools for users to experiment with. While human reviewers take a few days to return feedback, AI can offer immediate ideas for changes.

Generative AI can create unique text, images, video, and audio by analyzing existing content for patterns and reformulating it in a new way.

Critics worry that AI may replace human beings as the prime developers of creative content. But renowned music producer Rick Rubin thinks human creators will always be in demand. He says that while AI can create many random beats, people still require "the human curation aspect of art" to listen to those beats and decide which ones are worthy of being turned into songs.[1] In other words, good taste is a human trait that AI has not yet mastered.

PERSONAL WEB CONTENT

As the world of web content develops, traditional news sources such as telecasts and print newspapers will likely continue to decline. When companies began offering free online versions of their newspapers, they were unable to monetize the digital versions as much as they could with the print versions. A 2021 study by Pew Research found that between 2008 and 2020, the total number of US journalists fell by 26 percent. Meanwhile, the number of digital journalists grew.[2] These changes have continued over time, leaving an opening for web-based journalists.

With the rise of the internet, some news companies have begun firing reporters with decades of experience. To fill their roles, they've hired less-experienced writers and paid

PORTRAIT MODE

With more people watching video content on smartphones and tablets, experts predict a rise in videos that are created to be watched vertically. Studies suggest that most people hold their phones vertically when looking at the screen, even when watching videos. One sign of this shift is the trend for more vertical videos produced by professional content creators. Big-name brands such as the car company Mercedes are shooting commercials in portrait orientation, demonstrating the rising importance of vertical videos.

them less money. This has resulted in a dissolution of trust in traditional news media.

But some independent journalists are finding creative ways to work around this issue. Many are taking responsibility for reporting news within their own neighborhoods and professional fields. They do so by launching their own websites. Journalists develop their own brands and style, paying their own costs or enlisting help through crowdsourcing platforms such as Patreon and Kickstarter. This way, they can fund their investigations and production costs.

Creators of all genres of content will continue to monetize their work by connecting directly with their followers. Many platforms are available to help creators do this. These include Uscreen, Gumroad, Indiegogo, and Substack.

Uscreen caters to video creators, while Gumroad assists digital product creators. Indiegogo helps connect investors and startups. Substack allows short-form writers to monetize their work and build a devoted readership. Independent content creators are the new norm, and they will continue to see opportunities to build and engage with audiences online.

BLENDING REAL AND VIRTUAL WORLDS

The future of marketing content revolves around the concept of not disrupting the consumer's experience. While television commercials of the past interrupted entertainment to share messages about products, the newest advertising will attempt to blend into regular life. This could be made possible by the Internet of Things (IoT). This infrastructure of real-life objects is embedded with technology that connects to the internet to transmit information.

Smart glasses are another example. Technology companies such as Meta have developed high-tech eyeglasses that function much like a computer. They offer speakers, digital displays on the lenses, and a microphone for voice searches. Some developers are now working on contact lenses that can display direct messages and web content that is only visible to the wearer.

As technology bleeds into real life, the experience of being online becomes more realistic through extended reality (XR). This blend of immersive technologies, such as virtual reality (VR), works together to expand the possibilities of what people

THE FUTURE OF INFLUENCING

As the world of web content continues to evolve and change, the role of influencers will evolve too. According to the website Influencity, influencer marketing will likely become more data driven. Companies and organizations may begin using AI-based data analytics tools to accurately monitor the success of their influencer marketing tactics and predict outcomes. This development could affect which influencers businesses choose to partner with. The metaverse may also change the way influencers interact with their followers, opening up new digital worlds in which influencers can connect with audiences in an immersive, interactive way.

Meta partnered with eyewear brand Ray-Ban to create smart glasses. Each pair comes with a microphone, a camera, and other high-tech features. Customers can choose from more than 150 combinations of lenses and frames. 

can experience online. Inventors are designing wearable devices that allow users to realistically experience senses such as touch and sound while in VR.

Experts suggest that web content will grow so much more immersive that spending time online will feel like operating in the real world. Anything currently done in the real world might be recreated in a digital platform such as the metaverse. This online world was developed by the creators of Facebook. Avatars of individuals socialize, work, and enjoy the same things the real world has to offer. While Facebook simplified keeping in touch with distant friends and relatives, the metaverse promises to "get us even closer to that feeling of being together in person."[3]

At the same time, more screens and robots could possibly show up in physical stores and other real-life locations. This technology will give buyers the online shopping benefits they've become accustomed to, such as a wide selection of products and infinite options for customization. After years of moving information online, people are ready to use that electronic data to improve just about every aspect of human life.

As technology continues to change and expand the internet, the possibilities of web content are evolving too. But there is still room for people to create, share, and post things they care about. No matter what kind of web content people want to create, the future is sure to bring more and more opportunities for creative individuals.

MAKING WEB CONTENT PROFESSIONALLY

- A business, organization, or individual develops an idea for web content and defines their goals. This process may involve researching the market or audience and developing a specific web strategy.

- For production, some companies hire content creation teams made up of writers, UX specialists, web designers, or videographers. They may create website content, long or short videos, or social media posts.

- The postproduction process begins. The web content is edited and made ready for publication.

- The web content is posted online. The content may appear on a website or social media profiles, or both.

- The business, organization, or individual monitors the content and engages with their audience. They may use tools to analyze key performance indicators.

MAKING WEB CONTENT INDEPENDENTLY

- An individual or small team develops an idea for web content and defines their goals. This process may involve researching the target audience, developing a web strategy, and learning how to use computer programs. Beginners may need special equipment such as video cameras or microphones.

- The creator begins production, making content such as website posts, photos, long or short videos, or social media posts.

- The individual edits the web content and makes sure it is ready for publication. They may use video or audio editing software to do this.

- The individual posts or publishes the content on the web. The content may appear on the person's website, social media profile, or YouTube channel.

- The individual monitors the content and engages with the audience. They pay attention to what audiences like in order to plan future content.

QUOTE

"It's not the best content that wins. It's the best-promoted content that wins."

—*Andy Crestodina, chief marketing officer of Orbit Media*

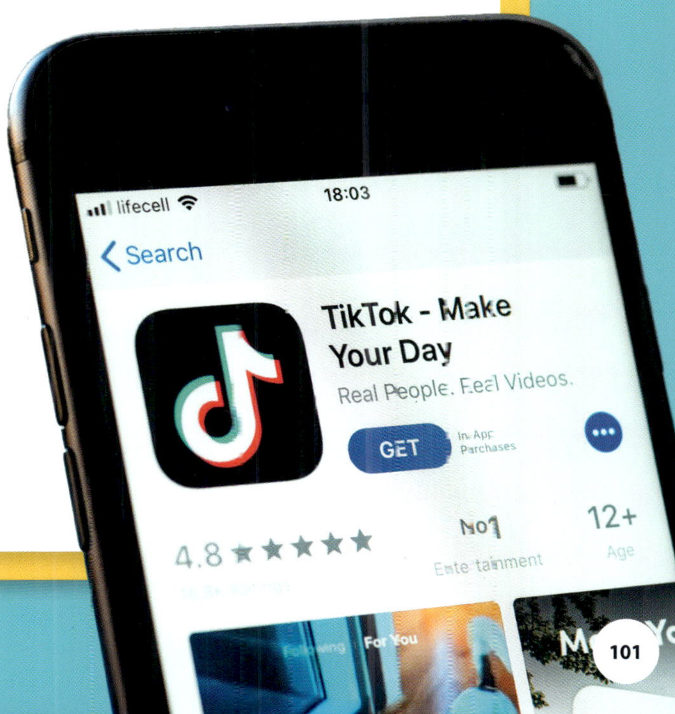

affiliate
Connected to or controlled by a group or business. Often used to describe a business arrangement in which a creator gets a bonus for every sale they generate for a business.

algorithm
A step-by-step set of instructions that tells a person or machine what to do.

cast
To play content from a phone or computer on a TV screen via a wireless connection.

commission
A sum of money paid to someone after they complete a task, such as selling goods or services.

database
An organized collection of content stored electronically.

entrepreneur
A person who takes risks in order to launch, manage, and operate a business.

freelance
Working independently by the hour, day, or project rather than working a long-term, salaried job for a single employer.

hypertext
Text that is linked to another document on the internet.

monetize
To use something to make money or to earn money from something.

non-fungible
Describing something that cannot be replaced or exchanged.

open-source
Of or relating to computer software that is freely available to use, study, edit, and modify.

patent
The exclusive right to an invention.

sync
A shorthand term for synchronize, or to match recorded audio with recorded footage after shooting.

transcribe
To convert spoken words into written words.

videographer
A person who records video productions with a camera.

SELECTED BIBLIOGRAPHY

Hennessy, Brittany. *Influencer: Building Your Personal Brand in the Age of Social Media.* Citadel, 2018.

Niederst Robbins, Jennifer. *Learning Web Design.* O'Reilly, 2018.

"The Video Marketing Playbook." *HubSpot*, n.d., offers.hubspot.com. Accessed 27 Dec. 2023.

FURTHER READINGS

Conley, Kate. *Social Media and Modern Society.* Abdo, 2022.

Edwards, Sue Bradford. *Making Podcasts.* Abdo, 2025.

Kallen, Stuart A. *Social Media's Star Power.* ReferencePoint, 2021.

ONLINE RESOURCES

To learn more about making web content, please visit **abdobooklinks.com** or scan this QR code. These links are routinely monitored and updated to provide the most current information available.

MORE INFORMATION

For more information on this subject, contact or visit the following organizations:

Internet Society
11710 Plaza America Dr., Ste. 400
Reston, VA 20190
internetsociety.org
The Internet Society works to extend the internet's reach and protect its
long-term well-being.

World Wide Web Consortium (W3C)
401 Edgewater Pl., Ste. 600
Wakefield, MA 01880
w3.org
The World Wide Web Consortium develops standards and guidelines to help everyone build
a web based on the principles of accessibility, internationalization, privacy, and security.

World Wide Web Foundation
1100 13th St. NW, Ste. 800
Washington, DC 20005
webfoundation.org
The World Wide Web Foundation uses research to discover digital inequalities around the
world. It works with governments to develop policies that increase access to the World
Wide Web.

CHAPTER 1. EVERYONE IS A CONTENT CREATOR

1. "How Many Websites Are There in the World?" *Siteefy*, 2 Dec. 2023, siteefy.com. Accessed 27 Dec. 2023.

2. Jack Flynn. "WordPress Market Share and Statistics [2023]: How Many Websites Use WordPress?" *Zippia*, 6 Feb. 2023, zippia.com. Accessed 27 Dec. 2023.

3. Claire Brotherton. "How to Start a Blog: What the Experts Recommend." *A Bright Clear Web*, 21 May 2021, abrightclearweb.com. Accessed 27 Dec. 2023.

CHAPTER 2. THE HISTORY OF WEB CONTENT

1. "What Was the First PC?" *Computer History Museum*, n.d., computerhistory.org. Accessed 27 Dec. 2023.

2. Tom Herbert and Alice Budd. "Sir Tim Berners-Lee: Net Worth, Quotes and Incredible Achievements of the World Wide Web Inventor." *Evening Standard*, 24 May 2023, standard.co.uk. Accessed 27 Dec. 2023.

3. Rani Molla. "How Apple's iPhone Changed the World: 10 Years in 10 Charts." *Vox*, 26 June 2017, vox.com. Accessed 27 Dec. 2023.

4. Neil C. Hughes. "How the iPhone Changed the World in 15 Years." *Cybernews*, 15 Nov. 2023, cybernews.com. Accessed 27 Dec. 2023.

5. Alexandra Samur and Colleen Christison. "The History of Social Media in 33 Key Moments." *Hootsuite*, 6 Apr. 2023, blog.hootsuite.com. Accessed 27 Dec. 2023.

6. Tamilore Oladipo. "23 Top Social Media Sites to Consider for Your Brand in 2024." *Buffer*, 20 Nov. 2023, buffer.com. Accessed 27 Dec. 2023.

7. Kyle Walsh and Jordan Malter. "Steve Jobs Explains the Original iPhone in 2007 CNBC Interview." *CNBC*, 25 Mar. 2019, cnbc.com. Accessed 27 Dec. 2023.

8. Kristi Hines. "The History of Social Media." *Search Engine Journal*, 2 Sept. 2022, searchenginejournal.com. Accessed 27 Dec. 2023.

CHAPTER 3. BUILDING A WEB PRESENCE

1. "How Many Websites Are There in the World? A Daily Calculator." *StatsFind*, 3 Jan. 2023, statsfind.com. Accessed 27 Dec. 2023.

2. "Social Media Fact Sheet." *Pew Research Center*, 7 Apr. 2021, pewresearch.org. Accessed 27 Dec. 2023.

3. "Teen Content Creators and Consumers." *Pew Research Center*, 2 Nov. 2005, pewresearch.org. Accessed 27 Dec. 2023.

4. "Teens, Social Media and Technology 2022." *Pew Research Center*, 10 Aug. 2022, pewresearch.org. Accessed 27 Dec. 2023.

5. Seth Godin. "The Most Important Page on the Web Is the Page You Build Yourself." *Seth's Blog*, 12 Dec. 2011, seths.blog. Accessed 27 Dec. 2023.

6. "Wengie YouTube Channel Statistics." *YouTubers.me*, n.d., us.youtubers.me. Accessed 27 Dec. 2023.

7. Shane Barker. "91 Experts Share the Most Effective SEO Tips to Drive Traffic to Your Website." *Medium*, 14 Feb. 2018, medium.com. Accessed 27 Dec. 2023.

CHAPTER 4. BUSINESS ON THE WEB

1. Jia Wertz. "The Growth of Subscription Commerce." *Forbes*, 15 July 2015, forbes.com. Accessed 27 Dec. 2023.

2. Ramona Sukhraj. "How Important Is a Top Listing in Google?" *Impact*, 12 Jan. 2017, impactplus.com. Accessed 27 Dec. 2023.

3. "The Video Marketing Playbook." *HubSpot*, n.d., offers.hubspot.com. Accessed 27 Dec. 2023.

4. Lian Klenk. "History of an Indispensable Tool: The Internet Movie Database (IMDb)." *TheatreArtLife*, 2 June 2022, theatreartlife.com. Accessed 27 Dec. 2023.

CHAPTER 5. CREATING VIDEO CONTENT

1. Maryam Mohsin. "10 Video Marketing Statistics That You Need to Know in 2023." *Oberlo*, 12 Jan. 2023, oberlo.com. Accessed 28 Dec. 2023.

2. Doug Ridley. "Top 3 Reasons Why Your Website Needs Videos." *Vital Design*, n.d., vitaldesign.com. Accessed 28 Dec. 2023.

3. "Why Are Short-Form Videos So Effective on Social?" *Wochit*, n.d., wochit.com. Accessed 28 Dec. 2023.

4. Emma Rose. "How Micro Video Can Supercharge Your Social Media Strategy." *IdeaRocket*, 31 Aug. 2022, idearocketanimation.com. Accessed 28 Dec. 2023.

5. Trishala Chokhani. "A Deep Dive into the Shrinking Attention Span." *eLearning Industry*, 23 Feb. 2023, elearningindustry.com. Accessed 28 Dec. 2023.

6. Meghan Stromberg. "YouTube Influencer Dave Amos Brings Urban Planning to the People." *American Planning Association*, 22 June 2023, planning.org. Accessed 28 Dec. 2023.

7. Phil Forbes. "The Anatomy of a Killer Unboxing Video: How and Why." *Packhelp*, n.d., packhelp com. Accessed 28 Dec. 2023.

8. "Marques Brownlee YouTube Channel Statistics." *YouTubers.me*, n.d., us.youtubers.me. Accessed 28 Dec. 2023.

9. "About YouTube." *YouTube*, n.d., about.youtube. Accessed 28 Dec. 2023.

10. Adam Wagner. "Are You Maximizing the Use of Video in Your Content Marketing Strategy?" *Forbes*, 15 May 2017, forbes.com. Accessed 28 Dec. 2023.

11. Brian Dean. "TikTok Statistics You Need to Know in 2024." *Backlinko*, 12 Dec. 2023, backlinko.com. Accessed 28 Dec. 2023.

CHAPTER 6. PROFESSIONAL ONLINE VIDEOS

1. Alana Semuels. "Inside the Making of CoComelon, the Children's Entertainment Juggernaut." *Time*, 15 Mar. 2022, time.com. Accessed 28 Dec. 2023.

2. "The Video Marketing Playbook." *HubSpot*, n.d., offers.hubspot.com. Accessed 27 Dec. 2023.

3. "History of TED." *TED*, n.d., ted.com. Accessed 28 Dec. 2023.

4. "TEDx Rules." *TED*, n.d., ted.com. Accessed 28 Dec. 2023.

5. Nahla Davies. "How Many Views Is Viral? Tips and Tricks on How to Go Viral." *Teachable*, n.d., teachable.com. Accessed 28 Dec. 2023.

6. Victor Luckerson. "Amazon to Buy Video Game Live-Streaming Site Twitch for $970 Million." *Time*, 25 Aug. 2014, time.com. Accessed 28 Dec. 2023.

7. "Team Up to Clean Up: TeamSeas." *Ocean Cleanup*, n.d., theoceancleanup.com. Accessed 28 Dec. 2023.

8. Jordan Miller. "37 Shopify Statistics for 2023: Facts and Figures for Merchants." *Gorgias*, 4 Oct. 2023, gorgias.com. Accessed 28 Dec. 2023.

CHAPTER 7. INFLUENCERS

1. Sarah Min. "86% of Young Americans Want to Become a Social Media Influencer." *CBS News*, 8 Nov. 2019, cbsnews.com. Accessed 28 Dec. 2023.

2. Sarah Israel. "Top Influencers in 2024: Who to Watch and Why They're Great." *Hootsuite*, 14 Feb. 2023, hootsuite.com. Accessed 28 Dec. 2023.

3. Lucas Shaw and Mark Bergen. "How a Kid Cracked YouTube's Secret Code." *Detroit News*, 1 Jan. 2021, detroitnews.com. Accessed 28 Dec. 2023.

4. Tanya Chen. "A Popular Teen YouTuber Broke Down in Tears While Recording a Blog about the Pressures of Being Perfect." *BuzzFeed News*, 7 Nov. 2019, buzzfeednews.com. Accessed 28 Dec. 2023.

5. Jeff Keleher. "9 Content Creation Quotes That Are Sure to Inspire." *Brafton*, 3 Oct. 2022, brafton.com. Accessed 28 Dec. 2023.

6. Colleen Christison. "How Much Do Influencers Make in 2024?" *Hootsuite*, 3 Oct. 2022, hootsuite.com. Accessed 28 Dec. 2023.

CHAPTER 8. THE FUTURE OF WEB CONTENT

1. Tim Ferriss. "The Tim Ferriss Show Transcripts: Rick Rubin." *Tim Ferriss*, 16 Jan. 2023, tim.blog. Accessed 28 Dec. 2023.

2. Mason Walker. "US Newsroom Employment Has Fallen 26% Since 2008." *Pew Research Center*, 13 July 2021, pewresearch.org. Accessed 28 Dec. 2023.

3. "What Is the Metaverse?" *Meta*, n.d., about.meta.com. Accessed 28 Dec. 2023.

4. Shirley Siluk. "Craig Davis on Interruption Marketing." *Collective Content*, 27 Sept. 2016, collectivecontent.agency. Accessed 28 Dec. 2023.

MARIE JASKULKA

When Marie Jaskulka started college at Saint Joseph's University in the 1990s, the school library had only one computer that could access the World Wide Web. Marie often monopolized it for hours at a time. When not writing books, Marie still spends a lot of time surfing the web. She writes poetry, fiction, and nonfiction and is the author of *The Lost Marble Notebook of Forgotten Girl & Random Boy*. She grew up in Philadelphia, Pennsylvania, and now lives in northeastern Pennsylvania.